GETTING STARTED WITH STATA
FOR WINDOWS®

A Stata Press Publication
STATA CORPORATION
College Station, Texas

Stata Press, 4905 Lakeway Drive, College Station, Texas 77845

The suggested citation for this software is

StataCorp. 2003. *Stata Statistical Software: Release 8.0*. College Station, TX: Stata Corporation.

Contents

Cross-Referencing the Documentation

When reading this manual, you will find references to other Stata manuals. For example,

[U]   **29 Overview of Stata estimation commands**
[P]   **matrix define**
[XT] **xtabond**

The first is a reference to Chapter 29, *Overview of Stata estimation commands* in the *Stata User's Guide*, the second is a reference to the `matrix define` entry in the *Stata Programming Reference Manual*, and the third is a reference to the `xtabond` entry in the *Stata Cross-Sectional Time-Series Reference Manual*.

All of the manuals in the Stata Documentation have a shorthand notation, such as [U] for the *User's Guide* and [P] for the *Stata Programming Reference Manual*.

The complete list of shorthand notations and manuals is as follows:

[GSM] *Getting Started with Stata for Macintosh*
[GSU] *Getting Started with Stata for Unix*
[GSW] *Getting Started with Stata for Windows*
[U]   *Stata User's Guide*
[R]   *Stata Base Reference Manual*
[G]   *Stata Graphics Reference Manual*
[P]   *Stata Programming Reference Manual*
[CL]   *Stata Cluster Analysis Reference Manual*
[XT]   *Stata Cross-Sectional Time-Series Reference Manual*
[SVY] *Stata Survey Data Reference Manual*
[ST]   *Stata Survival Analysis & Epidemiological Tables Reference Manual*
[TS]   *Stata Time-Series Reference Manual*

Detailed information about each of these manuals may be found online at

```
http://www.stata-press.com/manuals/
```

About this manual

Stata for Windows® is discussed in this manual. Stata for Macintosh users should see *Getting Started with Stata for Macintosh*; Stata for Unix users, see *Getting Started with Stata for Unix*.

This manual is intended both for people completely new to Stata and for experienced Stata users new to Stata for Windows. Previous Stata users will also find it helpful as a tutorial on some new features in Stata for Windows.

Each numbered chapter opens with a short summary of the contents of that chapter. New users can get a 15-minute course on Stata by reading all the summary pages first, and then going back and reading all or part of the text more thoroughly. Previous users will find the summaries a useful reference for Stata's basic commands.

Following the numbered chapters are four appendices with information specific to Stata for Windows.

We provide technical support to registered Stata users. Chapter 4 of this manual describes the sources of information available to help you learn about Stata's commands and features. One of these sources is the Stata web site (*http://www.stata.com*). Half of the web site is dedicated to user support. You will find answers to frequently asked questions (FAQs), as well as much other useful information for users. If, after looking at the Stata web site and the other sources of information described in Chapter 4, you still have questions, you can contact us as described in [U] **2.9 Technical support**.

1 Installation

Stata for Windows

Description Stata for Windows is available for all modern versions of Windows (95, 98, . . . , XP).
(Versions for Macintosh and for Unix are also available.)

Stata/SE Professional version of Stata.
Fastest.
Maximum of 32,766 variables; observations limited only by computer memory.
String variables up to 244 characters.
Matrices up to 11,000 x 11,000.

Intercooled Stata Professional version of Stata.
Very fast.
Maximum of 2,047 variables; observations limited only by computer memory.
String variables up to 80 characters.
Matrices up to 800 x 800.

Small Stata Stata for small computers.
Maximum of 99 variables and approximately 1,000 observations.
Slower than Stata/SE or Intercooled Stata.
String variables up to 80 characters.
Matrices up to 40 x 40.

You received **Whether you purchased Stata/SE, Intercooled Stata, or Small Stata:**
Single sheet *License and Authorization Key*.
Installation CD.
Registration Card.
A Base Stata Documentation Set includes:
Getting Started with Stata for Windows (this book).
Stata User's Guide.
4-volume *Stata Base Reference Manual*.
Stata Graphics Reference Manual.
A full Stata Documentation Set includes:
Getting Started with Stata for Windows (this book).
Stata User's Guide.
4-volume *Stata Base Reference Manual*.
Stata Graphics Reference Manual.
Stata Programming Reference Manual.
Stata Cluster Analysis Reference Manual.
Stata Cross-Sectional Time-Series Reference Manual.
Stata Survey Data Reference Manual.
Stata Survival Analysis & Epidemiological Tables Reference Manual.
Stata Time-Series Reference Manual.

Stata for Windows, continued

What to install   Look at your *License and Authorization Key*.

If you have a Stata/SE license:

Install Stata/SE.

If you have an Intercooled Stata license:

Install Intercooled Stata.
Do not install Stata/SE; your license will not
let you run it.

If you have a Small Stata license:

Install Small Stata.
Do not install Stata/SE or Intercooled Stata;
your license will not let you run them.

If you have a single-user license, you may install Stata on
both your work computer and your home computer, since it
would not be possible for you to use both at the same time.

Important   Do not lose your paper license. Keep it in a safe place.
You may need it again in the future.

Before you install

Before you begin the installation procedure:

1. Make sure that you have the Stata Installation CD.
2. Make sure that you have a License and Authorization Key.
3. Decide whether you are installing Stata/SE, Intercooled Stata, or Small Stata (see previous page).
4. Decide where you want to install the Stata software. We recommend C:\STATA8 or C:\STATA.
5. Decide where you want to set the working directory. This should be different from the installation directory so that files that you create will not get mixed up with Stata's files. We recommend C:\DATA.
6. If you already have an old version of Stata on your system, decide whether you want to keep it or delete it. If you want to install the new version in the same directory, you must let the installation program remove the old version. If you want to keep the old version, you must install the new version in a different directory.
7. You are ready to install; turn to the **Installation and upgrades for Windows** section on page 4.

Upgrade or update?

If you use Stata 7 or an earlier release, and you wish to upgrade to Stata 8, this is the chapter for you. If you have already installed Stata 8, and you wish to install the latest updates to Stata 8, see Chapter 20.

Installation and upgrades for Stata for Windows

Please be sure that Windows is installed and properly running before attempting to install Stata for Windows. Have your Stata License and Authorization Key with you.

1. Insert the CD in the CD-ROM drive.

2. If you have **Auto-insert Notification** enabled, the installer will automatically run. Otherwise, select **Run** from the **Start** menu and enter **D:\setup.exe** (assuming **D:** is the drive letter for your CD-ROM).

3. The Stata for Windows installation will begin. You will see a few opening screens of information. After reading these screens carefully, press the **Next** button.

4. The installation program will ask you where you want to install Stata.

    a. We recommend that you choose the default directory — C:\STATA8.

    b. When you have chosen an installation directory, press the **Next** button.

    c. If an older Stata is already installed in this directory, you will later be asked if you want to uninstall the older version. Only the official distribution files from the older version of Stata will be deleted in case you have some personal files in that directory. If you do not want to delete the previous version, you should choose a different directory in which to install Stata now.

Installation and upgrades for Stata for Windows, continued

5. At the **Select Components** dialog, you can choose which flavor of Stata to install.

a. Look at your License and Authorization Key. If it is a Small Stata license, press the **Small** button. Stata/SE and Intercooled Stata will not run under a Small Stata license.

b. If you have an Intercooled Stata license, press the **Intercooled** button. Stata/SE will not run under an Intercooled Stata license.

c. If you have a Stata/SE license, press the **Stata/SE** button.

6. The installation program will prompt you to press **Next** to begin the installation.

a. When you press **Next**, the installation program will determine if there is a previous version of Stata installed in the directory that you selected in step 4.

b. If there is a previous version, the installation program will ask you if it is okay to uninstall the previous version. Only the official distribution files from the older version of Stata will be deleted in case you have some personal files in that directory. If you do not want to delete the previous version, you should exit the installation, start over, and choose a different installation directory in step 4.

7. The installation program will then ask you to select the default working directory.

a. The default working directory is where your datasets, graphs, and other Stata-related files will be stored.

b. We recommend that you choose the default directory — C:\DATA.

c. When you have chosen a default working directory, press the **OK** button.

8. The installation program will copy Stata to your hard disk.

Installation and upgrades for Stata for Windows, continued

9. When the installation is complete, it will prompt you to press the **Finish** button to exit the installation.

10. The first time that you run Stata for Windows, it will prompt you for the information on your License and Authorization Key. Press *Tab* to move from one line to the next; press *Enter* or the **OK** button when finished. You must enter something for all fields in the dialog before you can continue. The code and authorization are not case-sensitive. If you make a mistake typing the codes, you will be prompted to try again.

```
Stata Initialization                          [X]

 [STATA icon]

   Name:            [                    ]

   Organization:    [                    ]

   Serial Number:   [                    ]

   Code:            [                    ]

   Authorization:   [                    ]

            [   OK   ]          [  Cancel  ]
```

11. You should now register your copy of Stata. You may do this either by completing the short online registration form at

   http://www.stata.com/register/

   or by filling out and mailing the registration card included with your software.

12. The installation is now complete. Turn to Chapter 2 for information on starting and stopping Stata for Windows.

2 Starting and stopping Stata

Testing the installation

Below we assume that Stata is installed. When you launch Stata from the Windows **Start** menu, the first screen you will see is

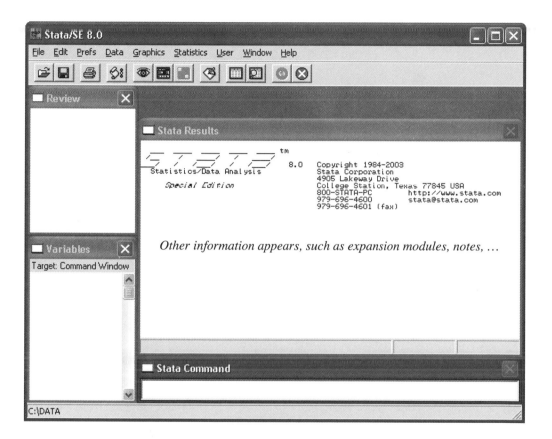

If you see a message box telling you that Stata could not find the license file, or if you receive any other error when you try to start Stata, please see [GSW] **B. Troubleshooting starting and stopping Stata** at the back of this book.

Note

There are other ways to start Stata than from the **Start** menu in Windows. You may find one of these ways more convenient. Please read [GSW] **A. More on starting and stopping Stata** at the end of this manual for more information.

The Stata windows

Past commands appear here   **Results are displayed here**

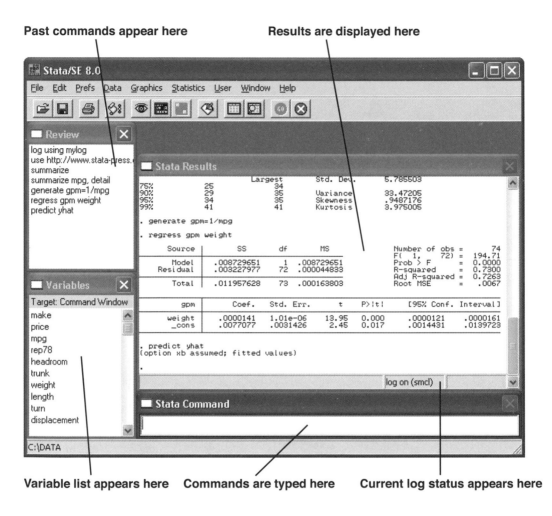

Variable list appears here   Commands are typed here   Current log status appears here

Note

If a window is open, but behind other windows, you can bring the window to the front of the other windows by selecting it from the **Window** menu. If you want to open a window, select the window from the **Window** menu.

The Stata toolbar

The Stata toolbar contains buttons that provide quick access to Stata's more commonly used features. If you ever forget what a button does, hold the mouse pointer over a button for a moment and a box will appear with a description of that button.

Open

> Open a Stata dataset.

Save

> Save to disk the Stata dataset currently in memory.

Print Graph/Print Viewer

> Print a graph or contents of Viewer.

Log Begin/Close/Suspend

> Begin a new log, append to an existing log, and close or suspend
> the current log. See Chapter 17 for an explanation of log files.

Viewer

> Open the Viewer or bring the Viewer to the front of the
> other Stata windows. See Chapter 3 for more information.

Bring Results to Front

> Bring the Results window to the front of the other Stata windows.

Continued

The Stata toolbar, continued

Bring Graph to Front

> Bring the Graph window to the front of the other Stata windows. See Chapter 16 for more information about graphs.

Do-file Editor

> Open the Do-file Editor or bring the Do-file Editor to the front of the other Stata windows. See Chapter 15 for more information.

Data Editor

> Open the Data Editor or bring the Data Editor to the front of the other Stata windows. See Chapter 6 for more information.

Data Browser

> Open the Data Browser or bring the Data Browser to the front of the other Stata windows. See Chapter 9 for more information.

Clear —more— Condition

> Tell Stata to continue when it has paused in the middle of long output. See Chapter 11 for more information.

Break

> Stop the current task in Stata. See Chapter 11 for more information.

Verifying the installation

Since you have just installed Stata for Windows, you should verify that the installation was performed correctly. Stata includes a command called `verinst`. Type `verinst` in the Command window and press *Enter*.

You should see output in the Results window informing you that Stata is correctly installed.

```
. verinst
You are running Stata/SE 8.0 for Windows.
Stata is correctly installed.
You can type exit to exit Stata.
```

If you see the error message below complaining that the command was not found,

```
. verinst
unrecognized command
r(199);
```

then not all of Stata is installed. Go back to Chapter 1.

If Stata informed you that it was correctly installed, you can do one of the following to exit:

1. Type `exit` in the Command window and press *Enter*. In the future, you may need to type `exit,` `clear` if your data has changed since you last saved it.

2. Choose **Exit** from the **File** menu.

3. Click on the close box (the box with an **X** in the upper right-hand corner of the Stata window).

The working directory

If you look at the screenshot on page 7, you will notice a status bar at the bottom of the screen that contains C:\DATA. Stata is telling you that C:\DATA is the current working directory.

The working directory is where graphs and datasets will be saved unless you specify another directory. See [GSW] **A.5 Starting Stata from other folders** for instructions on changing the default working directory.

Once you have started Stata, you can change the current working directory with the cd command. See [R] **cd** in the *Base Reference Manual* for full details.

Stata always displays the name of the working directory, so that it is easy to tell where your graphs and datasets will be saved.

Stata's interface

Stata is, at its heart, a command-driven application. That is, you issue commands to Stata to tell it to do things, such as load a dataset or perform a regression. As you become familiar with Stata's commands, you will find that they give Stata great power and flexibility.

You can type commands by hand (advanced users actually prefer this), but most of Stata's commands can also be accessed in a point-and-click manner by pulling down Stata's menus and selecting items that invoke dialog boxes to build Stata commands.

Stata's **Data**, **Graphics**, and **Statistics** menus provide point-and-click access to almost every command in Stata. As you will learn, Stata is fully programmable, and Stata programmers can even create their own dialogs and menus. The **User** menu provides a place for programmers to add their own menu items. Initially, it contains only some empty submenus; see [P] **window menu** for full details.

If you wish to perform a Poisson regression, you could type Stata's `poisson` command, or you could pull down the **Statistics** menu, open the submenu **Count outcomes**, and select **Poisson regression**, resulting in this dialog being displayed:

This dialog provides access to all of the functionality of Stata's `poisson` command. The `poisson` command has many options that can be accessed via multiple tabs across the top. The first time that you use the dialog box for a command, it is a good idea to look at the contents of each tab so that you will know everything of which the dialog box is capable.

The dialog boxes for many commands will have the **by/if/in** and **Weights** tabs. These provide access to Stata's ability to control the estimation sample and to deal with weighted data. Read [U] **14 Language syntax** for more information on these features of Stata's language.

Stata's interface, continued

Most dialog boxes in Stata will provide the same five buttons you see at the bottom of the Poisson dialog on the previous page: **OK**, **Cancel**, **Submit**, **?**, and **R**. **OK**, **Cancel**, and **Submit** do what you would expect.

Cancel dismisses the dialog box without doing anything. **OK** dismisses the dialog box and issues a Stata command based on how you have filled out the fields in the dialog box. **Submit** issues a command just like **OK**, but leaves the dialog box on the screen so that you can make changes and issue another command.

The command issued by a dialog box is submitted just as if you typed it by hand. You can see it in the Results window and the Review window after it executes; this will help you learn Stata's command syntax.

The button in the lower left corner of the dialog box with a question mark on it provides access to Stata's help system. Pressing it will typically take you to the help file for the Stata command associated with the dialog box. In this case, it would take you to the `poisson` help file.

The button with an **R** on it next to the help button is the "reset" button. Each time you open a dialog box, it will remember how you last filled it out. If you wish to reset its fields to their default values at any time, simply press this button.

In addition to being able to access the dialog boxes for Stata commands via Stata's menus, you can also invoke them via two other methods. You may know the name of a Stata command for which you want to see a dialog, but you may not remember how to navigate to that command in the menu system. Simply type db *commandname* to launch the dialog box for *commandname*:

```
. db poisson
```

You will also find access to the dialog box for a command in that command's help file; see Chapter 4 for more details.

As you read this manual, we will present examples of Stata commands. You may type those examples as presented, but you should also experiment with submitting those commands via their dialog boxes. Use the db command described above to quickly launch the dialog box for any command that you see in this manual.

Stata's interface, continued

Stata's Variables window can help you fill out the dialog boxes for Stata commands. When you open a dataset in Stata, the Variables window shows you a list of variable names in that dataset:

Notice the "Target" area at the top of the Variables window. When you click on a variable name in the Variables window, that name is typed into the specified target. By default, the Command window is the target, so if you click on a variable name, it will be typed into the Command window. You will read more about this in Chapter 10.

When you are filling out a dialog box to issue a Stata command, some fields in the dialog box may accept variable names. You can type the desired variable names in those fields, or you can take advantage of the Variables window. If you click in the **Dependent variable** field in the Poisson regression dialog box, the Variables window's target will become that field:

If you then click on a variable name in the Variables window, that name will appear in the **Dependent variable** field.

Increasing the amount of memory Stata uses

Stata loads your data into memory before analyzing it. By default, Stata/SE uses 10 megabytes of memory for data, Intercooled Stata uses 1 megabyte of memory for data, and Small Stata uses around 300 kilobytes of memory for data. Small Stata's memory area is fixed; you may ignore the rest of this page. Depending on the size of your datasets, you may wish to adjust the amount of memory Stata allocates for data if you are using Stata/SE or Intercooled Stata.

You can temporarily change the amount of memory Stata uses once it is running. You can also make that change permanent. For full details, see [GSW] **C. Setting the size of memory**.

To temporarily reset the amount of memory Stata/SE or Intercooled Stata is using for data, start Stata and type `set memory 5m`. The 5m means Stata is to allocate 5 megabytes to its data area. If you were to type 2m, Stata would allocate 2 megabytes. If you were to type 64m, Stata would allocate 64 megabytes. Change the number as you wish; Stata will tell you if your operating system is unable to provide the amount of memory that you request. Allocating too much memory to Stata may leave little memory available for other applications.

You can tell Stata to use a particular default memory allocation every time Stata is invoked. To do this, type `set memory 5m, permanently`. Stata will allocate 5 megabytes to its data area immediately, and it will remember that setting the next time you start it, since you told Stata to make that setting permanent.

Note

See [GSW] **A.8 Making shortcuts** and [GSW] **A.6 Specifying the amount of memory allocated** for more information on creating additional shortcuts to Stata with different amounts of memory. Also, see [GSW] **C. Setting the size of memory** for information on changing the amount of memory allocated for running Stata.

3 Using the Viewer

The Viewer

In this chapter, you will learn:

Stata has a Viewer in which you can
- View Stata help files and log files
- View Stata Markup and Control Language (SMCL) files
- View various other file types
- Print any of the above

Some Stata features invoke the Viewer:
- Viewing Stata logs
- Accessing Stata's help system

The Viewer has standard features:
- A **Back** button to return to the previous screen
- A **Refresh** button to refresh the current contents of the Viewer
- A **Search** button to open a Search dialog box
- A **Help!** button to obtain options for using the Viewer
- A **Contents** button to open the help system table of contents file
- A **What's New** button to list the new features of Stata
- A **News** button to list recent news and information of interest to Stata users
- A **Command** field to enter a variety of commands

To enter the Viewer and look at a SMCL file (such as a Stata log):
- Click on **View...** from the **File** menu
- Enter the *filename* in the field
- Or click on the **Browse...** button
  Select .smcl from the **Files of type** list
  Select the file you want to view
  Click on **Open**
- Click on **OK**

To view the current log in the Viewer:
- Select **Log** from the **File** menu
- Click on **View...**
- A dialog box will appear with the path and filename of the current log
- Click on **OK**
- The log that appears is a snapshot of the log at the time you opened the Viewer
- Click on the **Refresh** button at the top of the Viewer to see an updated snapshot of the current log

---

Continued

The Viewer, continued

To enter the Viewer and look at a local file:
- Click on **View...** from the **File** menu
- Enter the *filename* in the field
- Or press the **Browse...** button
  Select the appropriate file type from the **Files of type** list
  Select the file you want to view
  Click on **Open**
- Click on **OK**

To enter the Viewer and look at a file over the Internet:
- Click on **View...** from the **File** menu
- Enter the URL of the file in the field
- Click on **OK**

Once you are looking at a file, click on blue hypertext links to view different files:
- Blue text indicates a hypertext link
- Move your mouse over the blue text
- The link destination will be shown in the status bar at the bottom of the Viewer
- Click on the blue text to go to the link destination
- Click on the **Back** button at the top of the Viewer to go back to the previous screen

To print the contents of the Viewer:
- Select **Print Viewer...** from the **File** menu
- Or click on the **Print** button

Function of the Viewer

The Viewer is where you can see help information, look at (and print) logs of your current and previous Stata sessions, look at (and print) any other Stata formatted (SMCL) or plain text (ASCII) file, add new commands to Stata from the Internet, and install the latest official updates to Stata.

Note

> If you log output, your logs, by default, will be written in SMCL. SMCL stands for Stata Markup and Control Language. This is a language not unlike HTML. It contains directives specifying how the text is to be displayed. You can see [P] **smcl** for a complete description. If you are a programmer, you might want to exploit SMCL in the output your program produces.

> The Viewer understands SMCL, resulting in logs that appear in the Viewer just like they appeared originally in the Results window. Actually, your logs will look a little different because, by default, the Viewer has a white background and the Results window has a black background. You can, however, change the defaults of the Viewer to have a black background (which you could do by pulling down the **Prefs** menu and selecting **General Preferences...**) to make the log look exactly the same.

> You need to use the Viewer to print your SMCL logs because other programs do not understand SMCL. In addition, if you need to translate files containing SMCL to other formats, you can pull down the **File** menu, select **Log**, and click on **Translate...** (see [R] **translate** for more options).

The Viewer is part browser in that it may show links (shown in blue text) on which you can click to execute commands that the Viewer understands. If you are looking at a help file, a link might cause another help file to be shown in the Viewer. If you are looking at some sort of update screen, there might be a link where you can click to update your Stata. You can see to what command a link refers by moving the mouse on top of the link and looking in the status bar at the bottom of the Viewer.

These browser-like functions are an integral part of Stata and are not restricted to the Viewer. These functions apply to the Results window as well. The action of a link is the same regardless of the window in which it appears. If the action of a link is to show help on `logistic`, then clicking on that link will result in that same action no matter what window you are in. It will always display the help file in the Viewer.

Do not confuse the Viewer with the Results window. The Results window is where Stata output appears. The Viewer is a user-friendly window where you can click on links to look at help files and search results, look at your logs, install new Stata commands, and more.

Viewer buttons

Seven buttons appear at the top of the Viewer:

Back returns you to the previous screen viewed in the Viewer.

Refresh reloads the current contents of the Viewer.

Search allows you to perform a keyword search for help files, FAQs, *Stata Journal* articles and programs, STB articles and programs, and user-written programs.

Help! gives options for using the Viewer. It gives links, advice, and detailed instructions on how to use the Viewer. Especially valuable are its advice and examples for using **Search** and **Help**. Take a look at it.

Contents contains a table of contents for all the Stata help files, arranged by subject.

What's New describes the new features of Stata in the current release.

News lists recent news and information of interest to Stata users.

```
┌─────────────────────────────────────────────────────────────┐
│ ■ Stata Viewer [Help Contents]                          [X] │
├─────────────────────────────────────────────────────────────┤
│  Back │ Refresh │   Search   │  Help! │ Contents │What's New│ News │
├─────────────────────────────────────────────────────────────┤
│ Command: │help contents                                      │
├─────────────────────────────────────────────────────────────┤
│                                                               │
│  Top                                                          │
│  ─────────────────────────────────────────────────────────   │
│                                                               │
│  Category listings                                            │
│                                                               │
│      Basics                                                   │
│          language syntax, expressions and functions, ...      │
│                                                               │
│      Data management                                          │
│          inputting, editing, creating new variables, ...      │
│                                                               │
│      Statistics                                               │
│          summary statistics, tables, estimation, ...          │
│                                                               │
│      Graphics                                                 │
│          scatterplots, bar charts, ...                        │
│                                                               │
│      Programming & matrices                                   │
│          do-files, ado-files, and matrices                    │
│                                                               │
│  Help file listings                                           │
│                                                               │
│      Language syntax                                          │
│          advice on what to type                               │
│                                                               │
│      Copyrights                                               │
│  ─────────────────────────────────────────────────────────   │
│                                                               │
└─────────────────────────────────────────────────────────────┘
```

Uses of the Viewer

The Viewer provides a user-friendly interface to the various help functions of Stata. You can either click on the **Viewer** button or click on **Viewer** from the **Window** menu, and the Stata Viewer will open and provide you with links that allow you to perform a number of tasks.

In the Viewer, you can

- View the Contents of the help system
- View the help file for any Stata command
- Search Stata's documentation and FAQs by *keywords*
- Search net resources by *keywords*
- Obtain advice on how to use `search`
- Find and install *Stata Journal*, STB, and user-written programs
- Review, manage, and uninstall user-written programs
- Check for and optionally install official Stata updates
- View logs and other SMCL and ASCII files
- Launch a browser
- See the latest news from *http://www.stata.com/*

Viewing SMCL files

To view a Stata Markup and Control Language file (SMCL), such as a Stata log file, you will need to use the Viewer. To open a file and view its contents, simply click on **View...** from the **File** menu and you will be presented with a dialog box.

```
Choose File to View

File or URL:

[            ]

    OK       Cancel      Browse...
```

Type in the name of the file that you wish to view and click on **OK**. There is also a **Browse...** button that will allow you to search your hard drive for a file. To use this option, click on **Browse...**, select **SMCL Files (*.smcl)** from the **Files of type** list, select the file that you want to view, click on **Open**, and then click on **OK**. For example, suppose that you wish to view a SMCL log file, named `myfile.smcl`, which you have stored in `c:\data`.

1. *Click on* **View...** *from the* **File** *menu.*

2. *Click on the* **Browse...** *button.*

3. *Select* **SMCL Files (*.smcl)** *from the* **Files of type** *list.*

4. *Navigate the hard drive until the file is found.*

5. *Click once on the file.*

6. *Click on* **Open**.

7. *Click on* **OK**.

The SMCL file will now be displayed in the Viewer.

This is somewhat different from when you want to view the current log. In that case, you select **Log** from the **File** menu, click on **View...**, and the usual dialog box will appear, but with the path and filename of the current log already in the field. Simply click **OK**, and the log will appear in the Viewer. See Chapter 17 for more details.

Viewing other local files

You can view a variety of file types in the Viewer, such as plain text (ASCII) files. The process is very similar to viewing a SMCL file. The only difference is the file type that you will select from the **Files of type** list.

Viewing Internet files

If you want to look at a file over the Internet, the process is very similar to viewing a local text file, only instead of using the **Browse...** button, you simply type in the URL of the file that you want to see, such as *http://www.stata.com/man/readme.smcl*. Please note, however, that you do not want to just provide the URL for any web page. Web pages are written in HTML and are not translated by the Viewer. If you were to use the Viewer to look at a file over the Internet that was written in HTML, you would see the plain text HTML code of the document. You can launch your browser to view an HTML file by clicking on the browser link from the **Help!** screen, or by typing `browse` *URL* in the **Command** field at the top of the Viewer. Also note that you must provide the complete path of the file that you wish to view.

Links

Once you are viewing a file, you can click on the blue links to see related files or to execute commands that the Viewer understands. When you place the cursor over a blue link, the link destination will be displayed in the Viewer's status bar at the bottom of the window. After clicking on a link, you can click on the **Back** button to return to where you were.

Printing

To print the contents of the Viewer, select **Print Viewer...** from the File menu or just click on the **Print** button.

Commands in the Viewer

Everything done in the Viewer by clicking on links and buttons can also be done by issuing commands in the **Command** field at the top of the window. Some of the commands that can be issued in the Viewer are:

1. *Obtaining help*

    Type `help contents` to view the contents of Stata's help system.

    Type `help` *commandname* to view the help file for a Stata command.

2. *Searching*

    Type `search` *keyword* to search documentation and FAQs on a topic.

    Type `search` *keyword*`, net` to search net resources on a topic.

3. *User-written programs*

    Type `net from http://www.stata.com` to find and install *Stata Journal*, STB, and user-written programs from the net.

    Type `ado` to review user-written programs.

    Type `ado uninstall` to uninstall user-written programs.

4. *Updating*

    Type `update` to check on your current Stata version.

    Type `update query` to check on new official Stata update releases.

    Type `update all` to update your Stata.

5. *Viewing files*

    Type `view` *filename*`.smcl` to view SMCL files.

    Type `view` *filename*`.txt` to view ASCII files.

    Type `view` *filename*`.log` to view ASCII log files.

6. *Launching your browser to view an HTML file*

    Type `browse` *URL* to launch your browser.

7. *Keeping informed*

    Type `news` to see the latest news from *http://www.stata.com*.

4 Help

Online help

In this chapter, you will learn:

Stata for Windows has a comprehensive **Help** system.

You can have a help file open in the Viewer while you enter commands in the Command window.

When you choose **Help** from the main menu bar, you get a menu from which you can:
- See the help table of contents
- Search for help entries on a topic
- Get help for a Stata command
- List new features in Stata
- Read the latest Stata news
- Install the latest official updates from the Internet
- Install user-written, *Stata Journal*, or *Stata Technical Bulletin* programs for Stata from the Internet
- Go straight to important points on the Stata web site

Choosing **Search...** from the **Help** menu allows you to:
- **Search documentation and FAQs** by *keywords*
- **Search net resources** by *keywords*

Choosing **Search documentation and FAQs** from the **Search...** dialog box produces a screen containing:
- Hypertext links that will take you to the help files for the appropriate Stata commands
- References to the topic in the *Reference* manuals
- References to the topic in the *User's Guide*
- References to the topic in the *Stata Journal* and in the *Stata Technical Bulletin*
- Links to FAQs on Stata's web site dealing with the topic

Example:
- Select **Search...** from the **Help** menu
- Select **Search documentation and FAQs**
- Enter `regression` and click **OK**

You will get all the references to regression in the *Reference* manuals and the *User's Guide*, and a list of all the Stata commands that relate to regression:
- Stata commands like `areg`, `cnreg`, `cnsreg`, ... will appear in blue and are links

When you position the mouse pointer over a blue link, the pointer will change to a hand. The title of the linked file will appear in the status bar at the bottom of the Viewer. If you click while the hand is pointing at a command name like `areg`, you will go to the help file for `areg`.

Help, continued

Multiple topic words are allowed with **Search**
  Adding topic words will narrow the search; for example,
  - Enter `regression residuals`

Use proper English and statistical terminology when doing a **Search**.
  Do not enter the name of a Stata command when doing a **Search**.
  For example,
  - `t test` with **Search** is correct
  - `ttest` with **Search** will work, but in general, it is better to use proper English and statistical terminology with **Search**
  - `ttest` with **Stata command...** is correct because `ttest` is a Stata command
  - `help t test` in the Viewer edit field is incorrect
  - `help ttest` in the Viewer edit field is correct because `ttest` is a Stata command
  - `search t test` in the Viewer edit field is correct

Choosing **Search net resources** from the **Search...** dialog box produces
  a screen containing
  - Links that will take you to files on the Internet containing *Stata Journal*, STB, and user-written programs related to your *keywords*

Choosing **Contents** from the **Help** menu gives a listing of Stata's help
  table of contents.
  - You may choose from the links on this page to view help for a particular command, or
  - you may enter the full name of a Stata command in the edit field at the top of the Viewer

For example,
  - Click on the edit field at the top of the Viewer
  - Enter `help ttest` (`ttest` is a Stata command) and press *Enter*, and you will go to the help file for the Stata `ttest` command

The command help files also contain links.
  - Click on a link and you will go to another help file
  - Click on a dialog link and you will see the command dialog box

When you have stepped through a series of help files,
  - Click on **Back** to go back to the previous help file

You can start again at any time:
  - Just pull down **Help** from the main menu bar

Continued

Help, continued

The Viewer has six buttons other than **Back**:
    Refresh redisplays the current document.
    Search displays a dialog box where you may enter keywords
    to find help on a particular subject.

    Help! gives options on using the Viewer.

    Contents sends you to a subject table of contents with
    hypertext links for all help files.

    What's New describes the new features of Stata in the
    current release.

    News lists recent news and information from the Stata
    web site of interest to Stata users.

The help files contain lots of information, but
    not as much as the *Reference* manuals and the *User's Guide*.
    Help will tell you where to look in these manuals to find
    more information.

When we say "[U] **2.5 The Stata Journal and the Stata Technical Bulletin**", we mean
    section 2.5 in the *User's Guide*.

When we say "[R] **regress**",
    we mean the entry named **regress** in the *Base Reference Manual*.

When we say "[P] **#delimit**",
    we mean the entry named **#delimit** in the *Programming Reference Manual*.

For a complete list of manuals and their shorthand notations, see *Cross-Referencing
the Documentation*, which immediately follows the table of contents in this manual.

The *Stata Journal* provides
    • Articles on statistics, data management, graphics, and programming
    • New programs written by us (StataCorp) and by our users

The Stata web site (*http://www.stata.com*) provides
    • Answers to frequently asked questions (FAQs)
    • The latest official updates to Stata
    • Much, much more

The Help system

Let's illustrate the **Help** system with an example of how to use it.

1. *Choose* **Help** *from the main menu bar and select* **Search...**
2. *Select* **Search documentation and FAQs**.
3. *Enter* data *and click* **OK** *or press Enter.*

Keyword Search

⦿ Search documentation and FAQs

○ Search net resources

○ Search all

Keywords:

[   ]

[OK   ]   [Cancel]

The Help system, continued

4. *Stata will now search for all references to "data" among the Stata commands, the Reference manuals, the User's Guide, the Stata Journal, the Stata Technical Bulletin, and the FAQs on Stata's web site.* Here is the **Search** result:

```
■ Stata Viewer [search data]                                    X

  Back    Refresh      Search    Help!    Contents  What's New   News   ^

Command: search data

  GS        . . . . . . . . . . . . . . . . . . . . . . . Getting Started manual

  [U]       Chapter 9 . . . . . . . . . . . . . . . . . Stata's sample datasets

  [U]       Chapter 15 . . . . . . . . . . . . . . . . . . . . . . . . Data
            (help datatypes, missing, label, notes, format)

  [U]       Chapter 24 . . . . . . . . . . . . . . . . Commands to input data
            (help infiling)

  [U]       Chapter 25 . . . . . . . . . . . . . . . Commands for combining data
            (help append, joinby, merge)

  [U]       Chapter 27 . . . . . . . . . . . . . Commands for dealing with dates
            (help dates, tdates)

  [U]       Chapter 29 . . . . . . . . . . . Overview of Stata estimation commands
            (help estcom)

  [U]       Chapter 30 . . . . . . . . . . . . . Overview of survey estimation
            (help svy)

                                                                         v
```

The Help system, continued

5. *Scroll down until you see* [R] describe.
   `describe` is a Stata command that will "Describe contents of the dataset in memory or on disk".
   [R] means that the `describe` command is documented in the *Base Reference Manual.* The "help describe" means that there is an online help file for it.

```
Stata Viewer [search data]                                              ✕

Back  | Refresh  |    Search  |  Help!  | Contents | What's New |  News  | ʌ

Command: search data

    [R]     data types . . . . . . . . . . . . .  Quick reference for data types
            (help datatypes)

    [R]     describe . . . . . .  Describe contents of data in memory or on disk
            (help describe)

    [R]     destring . . . . . . . . . . . .  Change string variables to numeric
            (help destring)

    [R]     drawnorm . . . . . . . . . .  Draw a sample from a normal distribution
            (help drawnorm)

    [R]     drop . . . . . . . . . . . . . . . Eliminate variables or observations
            (help drop)

    [R]     duplicates . . . . . .  Inform on, tag, or drop duplicate observations
            (help duplicates)

    [R]     edit . . . . . . . . . . . . . . .  Edit & list data using Data Editor
            (help edit)

    [R]     egen . . . . . . . . . . . . . . . . . . . .  Extensions to generate
            (help egen)
                                                                             ᵥ
```

The Help system, continued

6. *Click on the link* "describe" *in* "help describe".

   Blue words in Stata are hypertext links. When you position the mouse pointer on or near a link, the pointer will change to a hand. When you click on the link, you will go to the corresponding help file, web page, etc. Clicking on the blue "describe" takes you to the help file for the `describe` command.

```
■ Stata Viewer [help describe]                                      ⊠
 ┌─────────┬──────────┬────────────┬─────────┬──────────┬───────────┬─────────┐  ^
 │  Back   │ Refresh  │   Search   │  Help!  │ Contents │ What's New│  News   │
 └─────────┴──────────┴────────────┴─────────┴──────────┴───────────┴─────────┘
 Command: help describe

 help for describe, ds, lookfor                     manual:  [R] describe
                                                    dialogs: describe varlist
                                                             describe using  ds

 Describe contents of data in memory or on disk

         describe [ varlist ] [, short detail fullnames numbers ]

         describe [ varlist ] using filename [, short detail varlist ]

         ds       [ varlist ] [, alpha varwidth(#) skip(#) ]

         lookfor  string [ string [...]]

 Description

     describe displays a summary of the contents of the data in memory or the
     data stored in a Stata-format dataset.

     ds lists variable names in a compact format.
                                                                              v
```

The Help system, continued

7. *Look at the help file for* describe.

   Help files for Stata commands contain (1) the command's syntax (i.e., the rules for how to type it), (2) links to dialogs, (3) a description of the command, (4) options, (5) examples, and (6) references to related commands. As you go through this manual, you will learn all about these things. For now, just scroll down to the examples. The examples tell us that it is valid to just type describe without anything else.

8. *Click on the Command window, type* describe, *and press Enter.*

   You may have to resize the Viewer so that you can see the bottom of the Results window. We see that describe tells us that we have no observations.

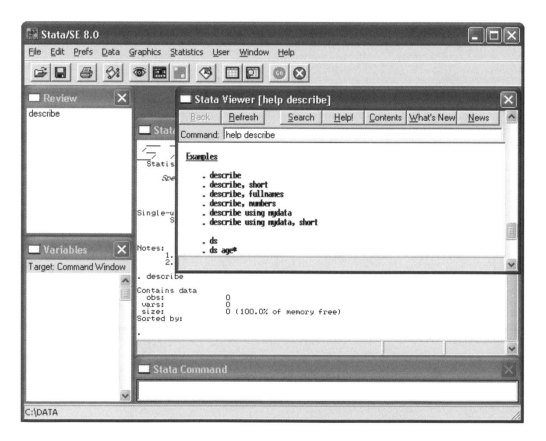

The Help system, continued

This example is the model of how we suggest you learn and use Stata. To find out which Stata command will perform the statistical or data management task you would like to do,

1. Choose **Help**, select **Search...**, select **Search documentation and FAQs**, and enter the topic.

2. Scan the **Search** results, click on the appropriate command name link, and go to its help file.

3. With the help file open in front of you, click on the Command window and enter the command, or click on one of the dialog links to open a dialog box for the command.

4. If the first help file you went to is not what you wanted, either look at the end of the help file for links to take you to related help files or click on **Back** to go back to the previous document and go from there to other help files.

5. If, at any time, you want to begin again with a new **Search**, click the **Search** button or select **Help** on the main menu bar and choose **Search...**.

6. If your **Search documentation and FAQs** search turned up no results, we suggest you look for *Stata Journal*, STB, and user-written programs on the Internet (materials available via Stata's `net` command) by selecting **Search net resources** and entering your topic again.

   If you select **Search documentation and FAQs**, Stata searches Stata's keyword database. If you select **Search net resources**, Stata searches for *Stata Journal*, STB, and user-written programs available for free download on the Internet; see Chapter 20 for more information.

7. Alternatively, you can select **Search all**. **Search all** is like selecting both **Search documentation and FAQs** and **Search net resources**.

Searching

If you enter `correlations` in the **Search** dialog box and select **Search documentation and FAQs**, Stata will search for all the sources of information that have to do with correlations.

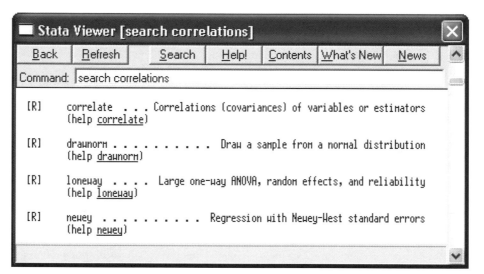

Search found several relevant entries in the *Base Reference Manual* and corresponding entries in the online help. If you are interested in correlations between variables, you could now click on `correlate` to get a brief summary of the feature, or you could look at [R] **correlate** in the *Base Reference Manual* for a complete description.

Search also reported some relevant entries from the *User's Guide*, the other *Reference* manuals, the *Stata Journal*, and the *Stata Technical Bulletin* (STB), but they are among the omitted output. See the end of this chapter for more information on *Stata Journal* and STB entries.

When you are using **Search**, use proper English and proper statistical terminology. Don't choose **Search...** and enter `ttest`. The term "ttest" is not proper statistical terminology. It is, however, the name of a Stata command. If you already know this and want to go directly to the help file for the `ttest` command, choose **Stata command...** from the **Help** menu, and type `ttest`. You can also type `help ttest` in the edit field at the top of the Viewer and press *Enter*.

Help needs to distinguish between topics and Stata commands since some of the names of Stata commands are general topics. For example, `logistic` is a Stata command. If you choose **Stata command...** and enter `logistic`, you will go right to the help file for the command. But if you choose **Search...** and enter `logistic`, you will get search results listing the many Stata commands that relate to logistic regression.

Search is designed to help you find information about statistics, graphics, data management, and programming features in Stata. When entering topics for the search, use appropriate terms from statistics, etc. For example, you could enter `Mann-Whitney`. Multiple topic words are allowed; e.g., `regression residuals`. For advice on using **Search**, click on the **Help!** button and click on **advice** in (**advice on using search**).

The Viewer buttons

Seven buttons appear at the top of the Viewer: **Back**, **Refresh**, **Search**, **Help!**, **Contents**, **What's New**, and **News**. These are covered in more detail in Chapter 3. A short summary is

Back returns you to the previous screen.

Refresh reloads the current screen.

Search opens a Search dialog box.

Help! gives options for using the Viewer.

Contents contains a table of contents for the help files.

What's New describes the new features of Stata.

News lists recent news and information of interest to Stata users.

Contents

If you click on the **Contents** button, you will see several help categories.

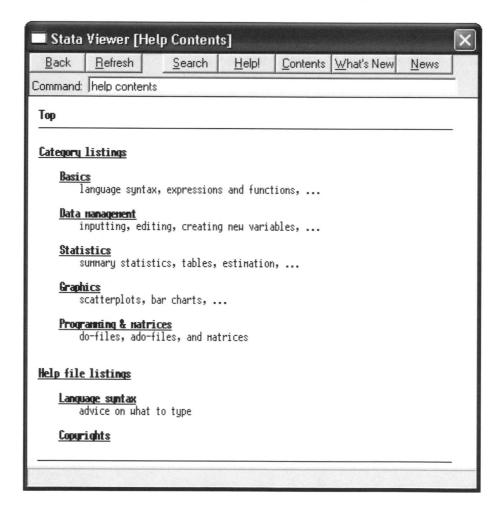

Suppose that we are interested in survival analysis. We click on Statistics.

Contents, continued

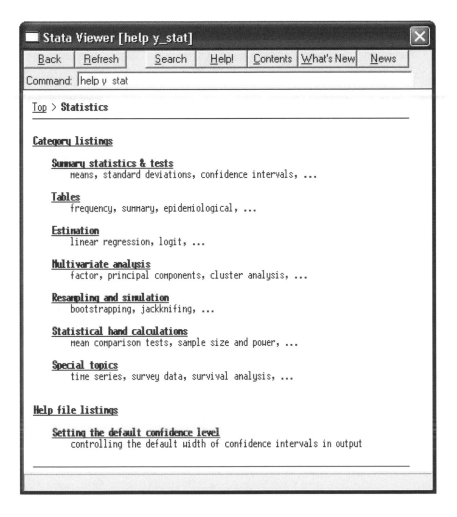

We select Estimation, although we could also have selected Special topics.

Contents, continued

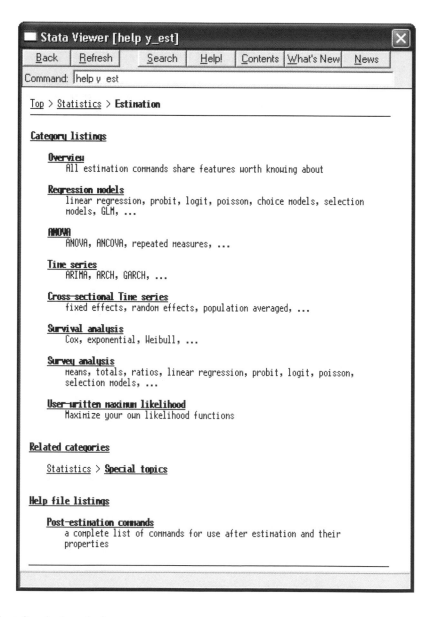

We select Survival analysis.

Contents, continued

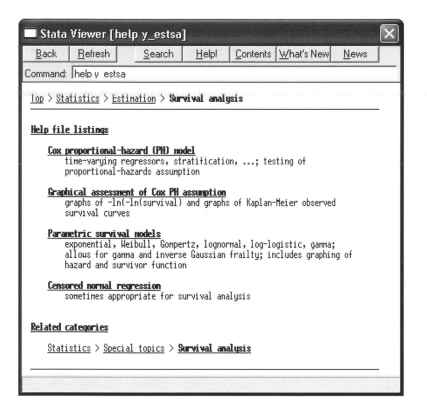

help and search commands

Notes

1. *You can also access Stata's help system from the Command window.*
   When you do so, the output can appear either in the Results window or in the Viewer.

2. *Typing* `search` *topic*
   in the Command window produces the same output as choosing **Search...** from the **Help** menu, selecting **Search documentation and FAQs**, and entering *topic*. However, the output will appear in the Results window.

3. *Typing* `search` *topic*`, net`
   in the Command window produces the same output as choosing **Search...** from the **Help** menu, selecting **Search net resources**, and entering *topic*. However, the output will appear in the Results window.

4. *Typing* `search` *topic*`, all`
   in the Command window produces the same output as choosing **Search...** from the **Help** menu, selecting **Search all**, and entering *topic*. However, the output will appear in the Results window.

5. *Typing* `help` *commandname*
   gives the same output as choosing **Stata command...** from the **Help** menu and entering *command-name*, except the output will appear in the Results window.

6. *Typing* `whelp` *commandname*
   is equivalent to choosing **Stata command...** from the **Help** menu and entering *commandname*. The help file will appear in Stata's Viewer.

7. See [U] **8 Stata's online help and search facilities** and [U] **8.8 search: All the details** in the *User's Guide* for more information about these command language versions of the **Help** system. The `search` command, in particular, has a few capabilities (such as author searches) that we have not demonstrated here.

The Stata Reference manuals and User's Guide

Things like [R] **ci**, [R] **hotelling**, and [R] **ttest** in the **Search** results and help files are references to the 4-volume *Stata Base Reference Manual*. You may also see things like [P] **#delimit**, which is a reference to the *Stata Programming Reference Manual*; and [U] **12 The Break key**, which is a reference to the *Stata User's Guide*. For a complete list of manuals and their shorthand notations, see *Cross-Referencing the Documentation*, which immediately follows the table of contents in this manual. Why bother to look in the manual when help is available in Stata? Because the manual has more information.

Stata's interactive help is extensive. If you zoomed through it at one second per line, it would take you several hours to read it all. However, although the online help might seem complete, it contains only about one-tenth of the information contained in the *Reference* manuals.

The Stata *Reference* manuals are each arranged like an encyclopedia—alphabetically. At the end of Volume 4 of the *Base Reference Manual* and at the end of the *User's Guide* is a combined index for the *User's Guide* and all the *Reference* manuals, except the *Graphics Reference Manual*. This combined index is a good place to start when you are looking for information about a command.

Entries are named things like **collapse**, **egen**, ..., **summarize**, which are generally themselves Stata commands.

For advice on how to use the *Reference* manuals, see Chapter 19 in this manual, or see [U] **1.1 Getting Started with Stata**.

The Stata Journal and the Stata Technical Bulletin

The **Search documentation and FAQs** facility searches all the Stata source materials, including the online help, the *User's Guide*, the *Reference* manuals, this manual, the *Stata Journal*, the *Stata Technical Bulletin* (STB), and the FAQs on Stata's web site.

The **Search net resources** facility searches all materials available via Stata's `net` command; see [R] **net**. These materials include the *Stata Journal*, STB, and user-written additions to Stata available on the Internet.

The *Stata Journal*, which began in 2001, is a quarterly journal containing articles about statistics, data analysis, teaching methods, and effective use of Stata's language. The journal publishes reviewed papers together with shorter notes and comments, regular columns, book reviews, and other material of interest to researchers applying statistics in a variety of disciplines. See *http://www.stata-journal.com* for additional information.

The predecessor to the *Stata Journal* is the *Stata Technical Bulletin* (STB). Even though the STB is no longer published, past issues contain articles and programs that may interest you. See *http://www.stata.com/bookstore/stbj.html* for the table of contents of past issues, and see the STB FAQ at *http://www.stata.com/support/faqs/res/stb.html* for detailed information on the STB.

Associated with each issue of both the *Stata Journal* and the STB are the programs and datasets described in the journals. These programs and datasets are made available for download and installation over the Internet, not only to subscribers, but to all Stata users. See [R] **net** and [R] **sj** for more information.

Since the *Stata Journal* and the STB have had a number of articles dealing with correlations, if we choose **Search...** from the **Help** menu, select **Search documentation and FAQs**, and enter `correlations`, we will see some of these references:

(*Continued on next page*)

The Stata Journal and the Stata Technical Bulletin, continued

```
┌──────────────────────────────────────────────────────────────┐
│ ■ Stata Viewer [search correlations]                       ⊠ │
├──────────────────────────────────────────────────────────────┤
│  Back  │ Refresh │    Search │  Help! │ Contents │What's New│ News │ ▲ │
├──────────────────────────────────────────────────────────────┤
│ Command: │search correlations                                │
├──────────────────────────────────────────────────────────────┤
│   SJ-2-1  st0009 . . . . . . . . . . . . . . . . . Transfer functions │
│           . . . . . . . . . . . . . . . . . . . . . . . . . A. McDowell │
│           Q1/02  SJ 2(1):70--84                    (no commands) │
│           demonstrates the estimation of a transfer function using │
│           Stata's arima command                                  │
│                                                                  │
│   SJ-1-1  st0004 Residual diagnostics for cross-section time series reg. models │
│           (help xttest2, xttest3 if installed) . . . . . . . . . . C. F. Baum │
│           Q4/01  SJ 1(1):101--104                                │
│           commands for testing groupwise heteroskedasticity and cross- │
│           sectional correlation after xtreg, fe and xtgls        │
│                                                                  │
│   STB-61  sg164 . . . . . . . Specification tests for linear panel data models │
│           (help xttest1 if installed) . . . . . W. Sosa-Escudero and A. K. Bera │
│           5/01   pp.18--21; STB Reprints Vol 10, pp.307--311     │
│           extension of xttest0 that computes seven specification tests │
│           for balanced error component models                    │
│                                                                  │
│   STB-61  snp15.3 . . . . . . . . . . . . . . . . . . Update to somersd │
│           (help somersd if installed) . . . . . . . . . . . . . . R. Newson │
│           5/01   p.22; STB Reprints Vol 10, p.324                │
│           updated for compatibility with Stata 7                 │
│                                                                  │
│   STB-59  sg159 . . . . . . . . . . . . . Confidence intervals for correlations │
│           (help ci2 if installed) . . . . . . . . . . . . . . . . P. T. Seed │
│           1/01   pp.27--28; STB Reprints Vol 10, pp.267--269     │
│           enhancement of ci and cii that provide confidence intervals │
│           for Pearson's product moment correlation and Spearman's │
│           rank correlation                                       │
│                                                              ▼  │
└──────────────────────────────────────────────────────────────┘
```

SJ-2-1 refers to Volume 2 Number 1 of the *Stata Journal*. Things like st0004, st0009, etc. are the way the *Stata Journal* refers to the articles that have appeared in it.

STB-59 refers to the fifty-ninth issue of the STB. Every six issues make a year's worth of STBs, and they are bound and republished in a volume called *Stata Technical Bulletin Reprints*. STB-59 could thus be found in *STB Reprints*, Volume 10. Things like sg159, snp15.3, etc. are the way the STB refers to the articles that have appeared in it.

In the screenshot above, you will notice that article st0004 in SJ-1-1 has two commands, `xttest2` and `xttest3`, associated with it. You can install these commands using Stata's Viewer. You may also install commands from STB articles via the Internet. See Chapter 20 for more information.

Links to other sites where you can freely download programs and datasets for Stata can be found on the Stata web site; see *http://www.stata.com/links/*. See Chapter 20 for more details on how to install this software. Also see [R] **ssc** for information on a convenient interface to resources available from the Statistical Software Components (SSC) archive.

We recommend that all users subscribe to the *Stata Journal*. See [U] **2.5 The Stata Journal and the Stata Technical Bulletin** for more information.

Notes

5 Loading and saving data

How to load your dataset from disk and save it to disk

In this one-page chapter, you will learn:

To load a Stata data file:
- Pull down the **File** menu and choose **Open...**
- Or type `use` *filename*
- Or click on the **Open** button

When loading a dataset, the dataset currently in memory is discarded.
Either save the current dataset first or allow it to be discarded:
- Choosing **File–Open...** will ask for your **OK** before discarding it
- Or type `use` *filename*`, clear`
- Or type `clear` and then `use` *filename*

To save a new dataset (or an old dataset under a new name):
- Pull down the **File** menu and choose **Save As...**
- Or type `save` *filename* in the Command window

To save a new dataset for use with Stata 7:
- Choose **Save As...** from the **File** menu and select **Stata 7 Data (*.dta)** from the **Save as type** list
- Or type `saveold` *filename* in the Command window

To resave a dataset that has been changed (overwriting the original data file):
- Pull down the **File** menu and choose **Save**
- Or click on the **Save** button
- Or type `save` *filename*`, replace`
- Or simply type `save, replace`

Stata data files are named *filename*`.dta`.
When saving or loading data files, the extension `.dta` is automatically added when an extension is not specified.

Datasets used in examples in the Stata manuals are available online.
Visit *http://www.stata-press.com/data/*

Important note: Changes are not permanent until you save them. You work with a copy of the dataset in memory, not with the data file itself.

Comment about resaving datasets: There is no way to recover your original data file once you have done a **File–Save** or `save, replace`. With important datasets, you may want either to keep a backup copy of your original *filename*`.dta`, or to save the changed dataset under a new name by choosing **File–Save As...** or by typing `save` *newfilename*.

Notes

6 Inputting data with the Data Editor

The Data Editor

In this chapter, you will learn:

To enter the editor:
- Click on the **Data Editor** button
- Or type `edit` and press *Enter* in the Command window

The editor is like a spreadsheet.
   Columns correspond to variables and rows to observations.
- You navigate by clicking on a cell or by using the arrow keys

You can copy and paste data between Stata's editor and other spreadsheets:
- Highlight the data that you wish to copy in either Stata or the other spreadsheet and then pull down **Edit** and choose **Copy**
- After copying the data, you can paste it into Stata's editor or into another spreadsheet by selecting the top left cell of the area to which you wish to paste, pulling down **Edit**, and choosing **Paste**

You modify or enter data by
- Choosing the cell, typing the value, and pressing *Enter* or *Tab*

The difference between *Enter* and *Tab*:
- *Enter* moves you down to the next cell in the column
- *Tab* moves you to the right to the next cell in the row (until you get to the last filled-in column, then it takes you back to the first column)

To input data variable-by-variable:
- Click on the top cell in the first empty column
- Enter the value
- Press *Enter* to go down

To input data observation-by-observation:
- Click on the first cell in the first empty row
- Type the value
- Press *Tab* to go right
- After the first observation's values have been entered, click on the second cell in the first column
- Type the values for the second observation; press *Tab* to go right
- At the end of the second observation (and all subsequent ones), *Tab* will automatically take you back to the first column

---

Continued

47

The Data Editor, continued

Numeric data and string data (i.e., data consisting of characters)
are entered the same way.
- You do not have to type double quotes around strings

Missing values for numeric variables are recorded as '.' (i.e., a period).
To enter a missing numeric value:
- Press *Enter* or *Tab*
- Or type '.' and press *Enter* or *Tab*
Stata supports different types of missing values if needed;
see [U] **15.2.1 Missing values** for more information

Missing values for string variables are just empty strings (i.e., nothing).
To enter a missing string value:
- Just press *Enter* or *Tab*

The editor initially names variables var1, var2, ...
You can rename them.

To rename a variable:
- Double-click anywhere in the variable's column
This brings up the **Variable Information** dialog:
- Enter the new name of the variable

A variable name must be 1 to 32 characters long.
- The characters can be letters: A – Z, a – z
- Or digits: 0 – 9
- Or underscores: _
- But no spaces or other characters
Example: My_Name1
The first character must be a letter or an underscore, but using
an underscore to begin your names is not recommended.

Stata is case-sensitive.
- Myvar, myvar, and MYVAR are different names

Inputting data

Suppose that we have the following dataset:

Make	Price	MPG	Weight	Gear Ratio
VW Rabbit	4697	25	1930	3.78
Olds 98	8814	21	4060	2.41
Chev. Monza	3667		2750	2.73
AMC Concord	4099	22	2930	3.58
Datsun 510	5079	24	2280	3.54
	5189	20	3280	2.93
Datsun 810	8129	21	2750	3.55

Note that we do not know MPG for the third car or the make of the sixth.

We will enter this dataset using the Data Editor.

1. *Call up the editor:*
   Click on the **Data Editor** button or type `edit` in the Command window.

Inputting data, continued

2. *Enter the data.*
   Data can be entered variable-by-variable or observation-by-observation. Columns correspond to variables and rows to observations.

3. *When entering data observation-by-observation, press Tab after each value.*
   To input data observation-by-observation, start in the top cell of the first column. Type the make VW Rabbit and hit *Tab* to go right. You do not want to hit *Enter* since it moves you down.

   Now enter the price 4697 and hit *Tab*. Continue until you have entered all the values for the first observation. Now click on the second cell in the first column, and enter the values for the second observation using the *Tab* key.

```
Stata Editor                                                      X
  Preserve  Restore    Sort      <<      >>      Hide    Delete...
                              var1[2] =
              var1       var2      var3      var4      var5
       1    VW Rabbit     4697       25      1930       3.78
```

4. *The Tab key is smart.*
   After the first observation has been entered, Stata knows how many variables you have. So, at the end of the second observation (and all subsequent observations), *Tab* will automatically take you back to the first column.

5. *When entering data variable-by-variable, press Enter after each value.*
   To enter data variable-by-variable, click on the top cell in the first empty column. Type the values for the variable and press *Enter* after each one.

Inputting data, continued

Things to know about entering data

1. *Quotes around strings are unnecessary.*
   For other Stata commands, you will learn that you must type double quotes (") around strings. You can use double quotes in the editor, too, but you do not have to bother.

2. *A period ('.') represents Stata's system missing numeric value.*

3. *Just press Tab or Enter to input a system missing numeric value.*
   MPG for the third observation is missing. To enter the missing value, just press *Tab* (or *Enter*), or type '.'.

4. *Just press Tab or Enter to input a missing value for a string variable.*
   We do not know the make of the sixth car, so just press *Tab* (or *Enter*), and there will be an empty string (i.e., nothing) for the value of make for this observation.

5. *Stata will not allow empty columns or rows in the middle of your dataset.*
   Whenever you enter new variables or observations, always begin in the first empty column or row. If you skip over some columns or rows, Stata will fill in the intervening columns or rows with missing values.

6. *When you see, for example,* var3[4]=
   This corresponds to the current cell that is highlighted. var3 is the default name of the third variable, and [4] indicates the fourth observation.

Stata Editor					
Preserve Restore Sort << >> Hide Delete...					
var3[4] = 22					

	var1	var2	var3	var4	var5
1	VW Rabbit	4697	25	1930	3.78
2	Olds 98	8814	21	4060	2.41
3	Chev. Monza	3667	.	2750	2.73
4	AMC Concord	4099	22	2930	3.58
5	Datsun 510	5079	24	2280	3.54
6		5189	20	3280	2.93
7	Datsun 810	8129	21	2750	3.55

Renaming variables

1. *The editor initially names variables* var1, var2, ..., var5 *when it creates them.*

2. *To rename a variable, double-click anywhere in the variable's column.*
   Double-clicking on any cell in the variable's column brings up the **Variable Information** dialog. Enter the new name of the variable. (You can also enter a variable label; see Chapter 8. You can also change the display format of the variable; see [U] **15.5 Formats: controlling how data are displayed**.)

   We will call the first variable make, the second price, the third mpg, the fourth weight, and the fifth gear_ratio. Just before we double-clicked and renamed var5 to gear_ratio, our screen looked like

Stata Editor								×
Preserve	Restore		Sort	<<	>>	Hide	Delete...	
			var5[3] = 2.73					
	make		price	mpg	weight	var5		
1	VW Rabbit		4697	25	1930	3.78		
2	Olds 98		8814	21	4060	2.41		
3	Chev. Monza		3667	.	2750	2.73		
4	AMC Concord		4099	22	2930	3.58		
5	Datsun 510		5079	24	2280	3.54		
6			5189	20	3280	2.93		
7	Datsun 810		8129	21	2750	3.55		

Rules for variable names

1. *Stata is case-sensitive.*
   Make, make, and MAKE are all different names to Stata. Had we called our variables Make, Price, MPG, etc., we would have to type them correctly capitalized in the future. Using all lowercase letters is easier.

2. *A variable name must be 1 to 32 characters long.*

3. *The characters can be letters* (A–Z, a–z), *digits* (0–9), *or underscores* (_).

4. *Spaces or other characters are not allowed.*

5. *The first character of a variable name must be a letter or an underscore.*
   Using an underscore to begin your names is not recommended, since Stata's built-in variables begin with an underscore.

Copying and pasting data

1. *Select the data that you wish to copy.*

   a. Click once on a variable name to select the entire column.

   b. Click once on an observation number to select the entire row.

   c. Click and drag the mouse to select a range of cells.

   As an example, we will copy and paste an observation from Stata's editor into itself. Highlight the observation by clicking on the observation number.

Stata Editor							
Preserve	Restore	Sort	<<	>>	Hide	Delete...	

make[1] = UW Rabbit

	make	price	mpg	weight	gear_ratio
1	UW Rabbit	4697	25	1930	3.78
2	Olds 98	8814	21	4060	2.41
3	Chev. Monza	3667	.	2750	2.73
4	AMC Concord	4099	22	2930	3.58
5	Datsun 510	5079	24	2280	3.54
6		5189	20	3280	2.93
7	Datsun 810	8129	21	2750	3.55

2. *Copy the data to the clipboard.*
   Pull down the **Edit** menu and choose **Copy**.

Copying and pasting data, continued

3. *Paste the data from the clipboard.*

   a. Click on the top left cell of the area to which you wish to paste.

	make	price	mpg	weight	gear_ratio
1	VW Rabbit	4697	25	1930	3.78
2	Olds 98	8814	21	4060	2.41
3	Chev. Monza	3667	.	2750	2.73
4	AMC Concord	4099	22	2930	3.58
5	Datsun 510	5079	24	2280	3.54
6		5189	20	3280	2.93
7	Datsun 810	8129	21	2750	3.55

Stata Editor — Preserve | Restore | Sort | << | >> | Hide | Delete... — make[8] =

   b. Pull down the **Edit** menu and choose **Paste**.

	make	price	mpg	weight	gear_ratio
1	VW Rabbit	4697	25	1930	3.78
2	Olds 98	8814	21	4060	2.41
3	Chev. Monza	3667	.	2750	2.73
4	AMC Concord	4099	22	2930	3.58
5	Datsun 510	5079	24	2280	3.54
6		5189	20	3280	2.93
7	Datsun 810	8129	21	2750	3.55
8	VW Rabbit	4697	25	1930	3.78

Stata Editor — Preserve | Restore | Sort | << | >> | Hide | Delete... — make[8] = VW Rabbit

Exiting the Data Editor

1. *To exit the editor:*
   Click on the editor's close box (the box with an **X** at the right of the editor's title bar).

2. *Changes made in the editor are not saved until you tell Stata to save them.*
   The data that you have entered only exist in computer memory. They are not yet saved on disk.

3. *Save your dataset by pulling down* **File** *and choosing* **Save As....**
   We will enter the filename `thecars` and Stata then adds the extension `.dta`, so the file is saved as `thecars.dta`. See the next chapter for full details on saving your dataset.

4. *Note You cannot save your dataset to disk until you exit the editor.*
   If you pull down the **File** menu while in the editor, **Save** and **Save As...** will be grayed out.

Important implication: Changes are not permanent until you save them to disk. In Stata, you work with data in memory, not with data on a disk file. You can tell Stata to make a temporary backup before making changes; read about the **Preserve** button in Chapter 9.

Fonts and the Data Editor

Want to use a different font in the editor? A larger font? A smaller font? While in the editor, click on the editor's control-menu box (click on the small box in the upper left corner of the window), choose **Font...**, and change it. Or, right-click anywhere in the editor and choose **Font...**.

Stata will remember the change even if you exit and restart Stata. To restore the default font, pull down **Prefs** from Stata's main menu bar and choose **Default Windowing**.

You can change the fonts of any of Stata's windows in this way. See Chapter 18 for details.

More on the Data Editor

The Data Editor also has a "browse" mode, which lets you safely look at your dataset without the possibility of accidentally changing it. See Chapter 9 for more information on the browse mode and other advanced features of the Data Editor.

Notes

7 Importing data

insheet, infile, infix, and odbc

In this chapter, you will learn:

Stata can import data from text (ASCII) files or from an ODBC source.	

There are 3 different commands for reading text files:	`insheet`
	 `infile`
	 `infix`

And one command for reading an ODBC source:	`odbc`
If the text file was created by a spreadsheet or a database program:	Use `insheet`.
If the data in the text file are separated by spaces and do not have string variables (i.e., nonnumeric characters), or if all strings are just one word, or if all strings are enclosed in quotes:	Use `infile`.
Otherwise, you will have to specify the format of the data in the text file:	Use `infix` or `infile` (fixed format).
If you need to read data from a database and that database supports ODBC:	Use `odbc`.
Spreadsheets save text files with delimiters:	`insheet` can read files delimited with commas or tab characters.
`insheet` cannot read space-delimited files:	Use `infile` as described on the next page.
`insheet` is easy to use:	It will read your variable names from the file, or it will make up variable names for you.
`insheet` is flexible:	You can override the variable names it would have chosen for you.
Before reading in data, you must clear memory. First save the current dataset (if you want it): Then clear memory by typing:	Choose **Save** from the **File** menu. `clear`

To read a text file created by a spreadsheet:
 To read myfile.raw:
 To read myfile.xyz:
 The extension .raw is assumed if you do not specify another extension.
	`insheet using` *filename*
	 `insheet using myfile`
	 `insheet using myfile.xyz`

Continued

insheet, infile, infix, and odbc continued

Suppose that you have data for three numeric variables
   separated by spaces entered in a text file.
   To read them in with the names a, b, and c, type: `infile a b c using` *filename*
   To read from file `myfile.raw`:   `infile a b c using myfile`
   To read from file `myfile.xyz`:   `infile a b c using myfile.xyz`
   The extension `.raw` is assumed if you
   do not specify another extension.

Variables are given the datatype `float`
   unless you specify otherwise: a, b, c will be `float`s.

Stata has six datatypes for variables:

   Real numbers, 8.5 digits of precision   `float`
   Real numbers, 16.5 digits of precision   `double`

   Integers between:
      −127 and 100   `byte`
      −32,767 and 32,740   `int`
      −2,147,483,647 and 2,147,483,620   `long`

   Strings (from 1 to 80 characters for Small Stata and
      Intercooled Stata; from 1 to 244 for Stata/SE)
      1-character long strings   `str1`
      2-character long strings   `str2`
      3-character long strings   `str3`

         ⋮   ⋮

      244-character long strings   `str244`

Continued

insheet, infile, infix, and odbc continued

The numeric variable types `byte` and `int` are used when you want to reduce the memory that your dataset requires:	See [U] **15 Data** and [U] **7 Setting the size of memory**.
The numeric variable types `long` and `double` are used for special accuracy requirements:	See [U] **16.10 Precision and problems therein**.
You specify the datatype by putting it in front of the variable name. If you put nothing in front, the variable will be a `float`.	
To `infile` 3 variables:	`infile a b c using myfile`
If string `a` is 10 characters or less and `b` and `c` numeric:	`infile str10 a b c using myfile`
If `a` is numeric, `b` a string, `c` numeric:	`infile a str10 b c using myfile`
If `a` and `b` are numeric, `c` a string:	`infile a b str10 c using myfile`
If `a` and `b` are strings, `c` numeric:	`infile str10 a str10 b c using myfile`
or:	`infile str10 (a b) c using myfile`
Missing values:	
Missing numbers are indicated by: . is called the system missing value, and .a, .b, ..., and .z are called extended missing values.	., .a, .b, ..., or .z
Missing strings are indicated by:	`""`
String values are in double quotes:	`"Harry Smith"`
If the string has no blanks, you can omit the quotes:	`Smith`
Stata can read formatted ASCII files. We give an example, but it is too complex to explain fully here:	See [R] **infile (fixed format)**, [R] **infix (fixed format)**, and [U] **24 Commands to input data**.
Other programs can be purchased that convert other software formats to Stata format:	See [U] **24.4 Transfer programs**.

insheet

The `insheet` command was specially developed to read in text (ASCII) files that were created by spreadsheet programs. Many people have data that they enter into their spreadsheet programs. All of the spreadsheet programs have an option to save the dataset as a text (ASCII) file. The spreadsheet programs will delimit the columns in the text files with either tab characters or commas. Additionally, the spreadsheet programs will sometimes save the column titles (variable names in Stata) in the text file.

In order to read in this file, you only have to type `insheet using` *filename*, where *filename* is the name of the text file. The `insheet` command will determine whether there are variable names in the file, what the separator character is (tab or comma), and what type of data are in each column. If *filename* contains spaces, put double quotes around the filename.

If you have a text (ASCII) file that you created by saving data from a spreadsheet program, then try the `insheet` command to read that data into Stata.

Remember that by default the `insheet` command understands files that use the tab or comma as the column delimiter. If you have a file that uses spaces as the delimiter, use the `infile` command instead. See the `infile` section of this chapter for more information. If you have a file that uses another character as the delimiter, you should use `insheet`'s `delimiter()` option; see [R] **insheet** for more information.

insheet, continued

Suppose that you have saved the file `sample.raw` from your favorite spreadsheet. It contains the following lines:

```
VW Rabbit,4697,25,1930,3.78
Olds 98,8814,21,4060,2.41
Chev. Monza,3667,,2750,2.73
,4099,22,2930,3.58
Datsun 510,5079,24,2280,3.54
Buick Regal,5189,20,3280,2.93
Datsun 810,8129,,2750,3.55
```

These correspond to the make, price, MPG, weight, and gear ratio of a few cars. Note that the variable names are not in the file (so `insheet` will assign its own names) and that the fields are separated by a comma. Use `insheet` to read the data:

```
. insheet using sample
(5 vars, 7 obs)

. list, separator(0)

              v1      v2    v3     v4      v5

  1.    VW Rabbit    4697    25    1930    3.78
  2.      Olds 98    8814    21    4060    2.41
  3.  Chev. Monza    3667     .    2750    2.73
  4.                 4099    22    2930    3.58
  5.   Datsun 510    5079    24    2280    3.54
  6.  Buick Regal    5189    20    3280    2.93
  7.   Datsun 810    8129     .    2750    3.55
```

If you want to specify better variable names, you can include the desired names in the command:

```
. insheet make price mpg weight gear_ratio using sample
(5 vars, 7 obs)

. list, separator(0)

             make    price    mpg    weight    gear_r~o

  1.    VW Rabbit     4697     25      1930        3.78
  2.      Olds 98     8814     21      4060        2.41
  3.  Chev. Monza     3667      .      2750        2.73
  4.                  4099     22      2930        3.58
  5.   Datsun 510     5079     24      2280        3.54
  6.  Buick Regal     5189     20      3280        2.93
  7.   Datsun 810     8129      .      2750        3.55
```

insheet, continued

Note

Stata listed gear_ratio as gear_r~o in the output from list. gear_r~o is a unique abbreviation for the variable gear_ratio. Stata displays the abbreviated variable name when variable names are longer than eight characters.

To prevent Stata from abbreviating gear_ratio, we could specify the abbreviate(10) option:

```
. list, separator(0) abbreviate(10)

             make    price   mpg   weight   gear_ratio

  1.     VW Rabbit    4697    25     1930         3.78
  2.       Olds 98    8814    21     4060         2.41
  3.   Chev. Monza    3667     .     2750         2.73
  4.                  4099    22     2930         3.58
  5.    Datsun 510    5079    24     2280         3.54
  6.   Buick Regal    5189    20     3280         2.93
  7.    Datsun 810    8129     .     2750         3.55
```

For more information on the ~ abbreviation and on list, see Chapter 11.

infile

The file `afewcars.raw` contains the following lines:

```
"VW Rabbit"       4697        25      1930    3.78
"Olds 98"    8814    21  4060    2.41
"Chev. Monza"     3667         .      2750    2.73
""                4099        22      2930    3.58
"Datsun 510"      5079        24      2280    3.54
"Buick Regal"
 5189
 20
 3280
 2.93
"Datsun 810"      8129       ***      2750    3.55
```

These correspond to the make, price, MPG, weight, and gear ratio of a few cars. Notice that the second line is not formatted the same as the first line; we do not know MPG on the third line (file contains '.'); we do not know make on the fourth line (file contains ""); the sixth car's data are spread over 5 lines; and we do not know MPG on the last line (file contains '***').

```
. infile str18 make price mpg weight gear_ratio using afewcars
'***' cannot be read as a number for mpg[7]
(7 observations read)
. list, separator(0)

                   make    price    mpg    weight    gear_r~o

        1.    VW Rabbit     4697     25      1930       3.78
        2.      Olds 98     8814     21      4060       2.41
        3.  Chev. Monza     3667      .      2750       2.73
        4.                  4099     22      2930       3.58
        5.   Datsun 510     5079     24      2280       3.54
        6.  Buick Regal     5189     20      3280       2.93
        7   Datsun 810     8129      .      2750       3.55

. save afewcars.dta, replace
```

Notes

1. `using afewcars` means using `afewcars.raw`. The extension `.raw` is assumed if you do not specify otherwise. If the data had been stored in `afewcars.asc`, we would have had to specify `using afewcars.asc`.

2. As in the Data Editor, '.' means numeric missing value.

3. As in the Data Editor, an empty string ("") means string missing value. However, with `infile` you must explicitly have a "" in your file so that it does not read the next number in the file as the value of this string variable.

4. Observations need not be on a single line. The location of line breaks does not matter.

5. Things that are not understood (such as ***) are mentioned and are stored as missing values.

infile with formatted data

File `cars2.raw` contains the following lines:

```
VW Rabbit          4697      25      1930    3.78
Olds 98            8814      21      4060    2.41
Chev. Monza        3667              2750    2.73
                   4099      22      2930    3.58
Datsun 510         5079      24      2280    3.54
Buick Regal        5189      20      3280    2.93
Datsun 810         8129              2750    3.55
```

These data are more difficult to read because (1) there are no double quotes around the strings in the first column and they include blanks, and (2) there are blanks in the first and third columns when we do not know the value.

`infile`'s ordinary logic when reading five variables is

1st thing	in the file is	1st variable	in 1st observation
2nd thing		2nd variable	in 1st observation
3rd thing		3rd variable	in 1st observation
4th thing		4th variable	in 1st observation
5th thing		5th variable	in 1st observation
6th thing		1st variable	in 2nd observation
7th thing		2nd variable	in 2nd observation
⋮		⋮	⋮

Carry out this logic on the above:

VW	is make	in 1st observation	
Rabbit	is price	in 1st observation	(error, stores as missing value)
4697	is MPG	in 1st observation	(wrong)
25	is weight	in 1st observation	(also wrong)
1930	is gear ratio	in 1st observation	(wrong again)
3.78	is make	in 2nd observation	(stores "3.78" as a string!)
Olds	is price	in 2nd observation	(error, stores as missing value)
98	is MPG	in 2nd observation	(surprise!)
⋮	⋮	⋮	

This problem is referred to as "loss of synchronization". There is a solution.

infile with formatted data, continued

Outside of Stata, using an editor or word processor, create `cars2.dct` containing the following lines:

```
dictionary using cars2.raw {
        _column(1)    str18 make      %11s
        _column(19)   price           %4f
        _column(31)   mpg             %2f
        _column(39)   weight          %4f
        _column(47)   gear_ratio      %4f
}
```

This is described in [R] **infile (fixed format)** in the *Base Reference Manual*. To read the data,

```
. infile using cars2
dictionary using cars2.raw {
        _column(1)    str18 make      %11s
        _column(19)   price           %4f
        _column(31)   mpg             %2f
        _column(39)   weight          %4f
        _column(47)   gear_ratio      %4f
}
(7 observations read)
. list

                   make    price   mpg   weight   gear_r~o

        1.      VW Rabbit    4697    25     1930       3.78
        2.        Olds 98    8814    21     4060       2.41
        3.    Chev. Monza    3667     .     2750       2.73
        4.                   4099    22     2930       3.58
        5.      Datsun 510   5079    24     2280       3.54

        6.    Buick Regal    5189    20     3280       2.93
        7.     Datsun 810    8129     .     2750       3.55
```

That is,

1. Create another file describing the contents of the dataset with another program — an editor or word processor. Name the file *myfile*.`dct`, and save it as a text (ASCII) file. Note that editors such as Notepad and WordPad may add a `.txt` extension to the file, resulting in *myfile*.`dct`.`txt`. To prevent this, enclose the filename in double quotes (") when you save it from an editor.

2. Instructions for creating the contents of this file can be found in [R] **infile (fixed format)**.

3. Once the dictionary exists, you type `infile using` *myfile*. The file extension `.dct` is assumed when `infile` is used without any variable names.

Importing data from ODBC sources

ODBC, an acronym for Open DataBase Connectivity, is a standard for exchanging data between programs. Stata supports the ODBC standard for importing data via the odbc command and is capable of reading from any ODBC data source on your computer.

If you have not heard of ODBC before, then you do not need to use Stata's odbc command; continue with the next section in this chapter.

The easiest way to start using odbc is to type odbc list:

```
. odbc list
Data Source Name                  Driver
MS Access Database                Microsoft Access Driver (*.mdb)
Excel Files                       Microsoft Excel Driver (*.xls)
Visual FoxPro Database            Microsoft Visual FoxPro Driver
Visual FoxPro Tables              Microsoft Visual FoxPro Driver
dBase Files - Word                Microsoft dBase VFP Driver (*.dbf)
FoxPro Files - Word               Microsoft FoxPro VFP Driver (*.dbf)
dBASE Files                       Microsoft dBase Driver (*.dbf)
```

What you see will be different depending on what ODBC sources are available on your computer. After seeing what data sources are available, you can then choose to import data from tables within a particular data source. The full details go beyond the scope of this manual; read [R] **odbc** for complete instructions.

Importing files from other software

If you have a file created by another software package that you would like to read into Stata, you should be able to use options on that software package to output the file as a plain text (ASCII) file. Then, you can use `infile`, `insheet`, or `infix` to read in the data.

Or, you may be able to copy and paste data from another spreadsheet into Stata's Data Editor. See [R] **edit** for full details.

Or, you can purchase a transfer program that will convert the other software's data file format to Stata's data file format. See [U] **24.4 Transfer programs**.

Notes

8 Labeling data

describe and label

In this chapter, you will learn:

To obtain a description of the data
    currently in memory: `describe`
    stored on disk: `describe using` *filename*

To set or reset the data label: `label data "`*text*`"`

To set or reset a variable's label: `label var` *varname* `"`*text*`"`

To label a variable's values:
    Create a value label: `label define` *lblname* `#` `"`*text*`"` `#` `"`*text*`"` ...
    Attach value label to variable: `label values` *varname* *lblname*

    You can attach the same value
    label to other variables, too.

To unlabel
    the data: `label data`
    a variable: `label var` *varname*
    a variable's values: `label values` *varname*

To delete a value label: `label drop` *lblname*

To change a value label:
    Delete it: `label drop` *lblname*
    Redefine it: `label define` *lblname* `#` `"`*text*`"` `#` `"`*text*`"` ...

To make changes permanent,
    resave the dataset: Choose **Save** under the **File** menu
    or, alternatively, you can type: `save` *filename*`,` `replace`

Note

We cover only the rudiments of value labels here. For instance, there is a way to modify value labels without deleting and redefining. See [U] **15.6.3 Value labels** for a complete treatment.

describe

In Chapter 7, we saved a dataset called `afewcars.dta`:

```
. use afewcars

. list, separator(0)

            make    price    mpg    weight    gear_r~o

  1.    VW Rabbit     4697     25      1930        3.78
  2.      Olds 98     8814     21      4060        2.41
  3.   Chev. Monza    3667      .      2750        2.73
  4.                  4099     22      2930        3.58
  5.    Datsun 510    5079     24      2280        3.54
  6.   Buick Regal    5189     20      3280        2.93
  7.    Datsun 810    8129      .      2750        3.55
```

The description of this dataset is

```
. describe
Contains data from afewcars.dta
  obs:            7
  vars:           5                              9 Sep 2002 14:25
  size:         266 (99.9% of memory free)

              storage  display      value
variable name   type   format       label      variable label

make           str18   %18s
price          float   %9.0g
mpg            float   %9.0g
weight         float   %9.0g
gear_ratio     float   %9.0g

Sorted by:
```

1. The variable name is how we refer to the column of data.

2. The storage type was described in Chapter 7.

3. The display format is described in [U] **15.5 Formats: controlling how data are displayed**. It controls how the variable is displayed. By default, Stata sets it to something reasonable given the storage type.

Describing a file without loading it

It is not necessary to load a dataset before describing it:

```
. describe using afewcars
Contains data
  obs:            7                            9 Sep 2002 14:25
 vars:            5
 size:          266

              storage  display    value
variable name   type   format     label      variable label

make            str18  %18s
price           float  %9.0g
mpg             float  %9.0g
weight          float  %9.0g
gear_ratio      float  %9.0g

Sorted by:
```

That is,

1. When you type describe by itself, Stata describes the current contents of memory.

2. When you type describe using *filename*, Stata describes the contents of the specified Stata data file (i.e., a file named *filename*.dta created by Stata).

describe and labels

Datasets can contain labels on the dataset, variables, and values. As an example, we use the dataset auto.dta, which is stored in the directory where you installed Stata. It is also stored online at the Stata Press web site, and Stata can even read it from that location! You will learn more about Stata's Internet capabilities in Chapter 20.

```
. describe using http://www.stata-press.com/data/r8/auto
Contains data from http://www.stata-press.com/data/r8/auto.dta
  obs:            74                          1978 Automobile Data
  vars:           12                          14 Oct 2002 09:02
  size:        3,478                          (_dta has notes)

              storage  display    value
variable name  type    format     label    variable label

make          str18   %-18s                Make and Model
price         int     %8.0gc               Price
mpg           int     %8.0g                Mileage (mpg)
rep78         int     %8.0g                Repair Record 1978
headroom      float   %6.1f                Headroom (in.)
trunk         int     %8.0g                Trunk space (cu. ft.)
weight        int     %8.0gc               Weight (lbs.)
length        int     %8.0g                Length (in.)
turn          int     %8.0g                Turn Circle (ft.)
displacement  int     %8.0g                Displacement (cu. in.)
gear_ratio    float   %6.2f                Gear Ratio
foreign       byte    %8.0g     origin     Car type

Sorted by:  foreign
```

That is,

variable name		display format		variable label
↓		↙		↙
gear_ratio	float	%6.2f		Gear Ratio
foreign	byte	%8.0g	origin	Car type
	↑		↑	
	storage type		value label name	

1. Value labels allow numeric variables — such as foreign — to have words associated with numeric codes. The describe tells us that the numeric variable foreign has value label origin associated with it. Although not revealed by describe, the variable foreign takes on the values 0 and 1, and the value label origin associates 0 ↔ Domestic and 1 ↔ Foreign. When we browse the data (see Chapter 9), it appears as if foreign contains the values "Domestic" and "Foreign". Note that it is not necessary for the value label to have a name different than the variable. We could just as easily have used a value label named foreign.

2. Variable labels are merely comments that help us remember the contents of variables. They are used by many Stata commands to enhance tabular and graphical output.

Labeling datasets and variables

```
. use afewcars
. describe
Contains data from afewcars.dta
  obs:              7
  vars:             5                              9 Sep 2002 14:25
  size:           266 (99.9% of memory free)

              storage  display    value
variable name  type    format     label      variable label

make          str18    %18s
price         float    %9.0g
mpg           float    %9.0g
weight        float    %9.0g
gear_ratio    float    %9.0g

Sorted by:
. label var make "Make of car"
. label var gear_ratio "Gear ratio"
. label data "A few 1978 cars"
. describe
Contains data from afewcars.dta
  obs:              7                              A few 1978 cars
  vars:             5                              9 Sep 2002 14:25
  size:           266 (99.9% of memory free)

              storage  display    value
variable name  type    format     label      variable label

make          str18    %18s                   Make of car
price         float    %9.0g
mpg           float    %9.0g
weight        float    %9.0g
gear_ratio    float    %9.0g                  Gear ratio

Sorted by:
. save afewcars, replace
file afewcars.dta saved
```

That is,

1. Use `label var` to set or reset a variable's label. Put the label itself in double quotes:

    `label var gear_ratio "Gear Ratio"`

2. Use `label data` to set or reset the dataset's label.

3. You set or change the labels on the dataset in memory. To make the changes permanent, resave the dataset.

Labeling values of variables

We have secretly added the variable `foreign` to `afewcars.dta`:

```
. list, separator(0)

              make   price   mpg   weight   gear_r~o   foreign

   1.     VW Rabbit    4697    25     1930       3.78         1
   2.       Olds 98    8814    21     4060       2.41         0
   3.   Chev. Monza    3667     .     2750       2.73         0
   4.                  4099    22     2930       3.58         0
   5.     Datsun 510   5079    24     2280       3.54         1
   6.   Buick Regal    5189    20     3280       2.93         0
   7.    Datsun 810    8129     .     2750       3.55         1
```

Labeling values of variables, continued

`foreign` $= 0$ denotes domestic cars and 1 denotes foreign. Label the values "domestic" and "foreign":

```
. label define origin 0 "domestic" 1 "foreign"
. label values foreign origin
. describe
Contains data from afewcars.dta
  obs:             7                          A few 1978 cars
  vars:            6                          9 Sep 2002 14:25
  size:          294 (99.8% of memory free)

              storage  display    value
variable name  type    format     label      variable label

make           str18   %18s                   Make of car
price          float   %9.0g
mpg            float   %9.0g
weight         float   %9.0g
gear_ratio     float   %9.0g                  Gear ratio
foreign        float   %9.0g     origin

Sorted by:
     Note:  dataset has changed since last saved
. save afewcars, replace
file afewcars.dta saved
```

That is,

1. Use `label define` to create a value label. The syntax is

    `label define` *labelname* *#* "*contents*" *#* "*contents*" ...

2. Use `label values` to associate the label with a variable. The syntax is

    `label values` *variablename labelname*

3. To make the change permanent, resave the dataset.

Notes

9 Changing and viewing data with the Data Editor

Advanced use of the Data Editor

In this chapter, you will learn:

You can select the variables that appear in the editor:
Type in the Command window:
- `edit make`   Selects the single variable `make`
- `edit make mpg`   Selects the variables `make` and `mpg`

You can include any number of variables.
- See Chapter 10 for shortcuts for entering variable names.

You can restrict the observations that appear in the editor:
Type in the Command window:
- `edit in 1`   Uses only the first observation
- `edit in 2`   Uses only the second observation
- `edit in -2`   Uses only the second from the last observation
- `edit in -1`   Uses only the last observation
- `edit in l`   (i.e., ℓ)   Also uses only the last observation

To restrict to a series of observations:
- `edit in 1/9`   Uses observations 1 through 9
- `edit in 2/-2`   Uses observations 2 through second from the last

To restrict to those observations satisfying a mathematical expression:
- `edit if` *exp*   Uses observations for which *exp* is true
- `edit if mpg>20`   Uses observations for which `mpg>20`
- `edit if mpg==20`   Uses observations for which `mpg` equals 20
- `edit if missing(rep78)` Uses observations for which `rep78` equals missing

You can combine `in` and `if` (the order does not matter):
- `edit in 1/9 if mpg>=25`
- `edit if price<6000 in 5/-1`

You can select variables and restrict observations at the same time:
- `edit make in 5/-5`
- `edit make mpg if mpg>=25`
- `edit make mpg if missing(rep78)`
- `edit make mpg in 1/9 if mpg>=25`

Continued

77

Advanced use of the Data Editor, continued

To make changes to your data:
- Type `edit` alone or `edit` *varname*(*s*), `edit if` ..., etc.
- Or click on the Data Editor button (but you cannot select variables or restrict observations if you enter the editor this way)
- Once in the editor, click on a cell that you want to change
- Enter a new value and press *Enter* or *Tab*

Restricting the editor to certain variables and observations lessens the chance of making a serious mistake. For example, if you need to change `mpg` when it is missing, you can limit the data's exposure by typing
- `edit make mpg if missing(mpg)`

Note: For making systematic global changes to your data, the `replace` command may be more appropriate.
- See Chapter 12

To delete variables and observations:
- Press the **Delete...** button; see the next page

Note: For deleting groups of observations or variables, the `drop` command is more appropriate.
- See Chapter 13

The Data Editor can be used to view data.
  To enter the editor in `browse` mode:
- Click on the **Data Browser** button
- Or type `browse` in the Command window

The `browse` mode is useful since you cannot accidentally change your data. Use `browse`, not `edit`, when you just want to look.

You can also select variables and restrict observations with `browse`.
  Examples:
- `browse make mpg`
- `browse in 1/20`
- `browse if missing(rep78)`
- `browse make mpg rep78 in 5/-5 if mpg>20`

In later chapters, you will learn:

What is true for `edit` and `browse` is true for almost all Stata commands: Type the command, optionally followed by a variable list, optionally followed by an `if`, optionally followed by an `in`.

Buttons on the Data Editor

The editor has seven buttons:

Preserve
> If you make changes to your data in the Data Editor and
> are satisfied with them (and plan to stay in the editor to
> make additional changes), you can update the backup copy:
> - Click on **Preserve**

Restore
> Stata automatically makes a backup copy of your data
> when you enter the editor.
>
> If you want to cancel the changes that you made and
> restore the backup copy:
> - Click on **Restore**

Sort
> **Sort** sorts the observations of the current variable in ascending order

<<
> The **<<** button shifts the current variable to be the first variable.

>>
> The **>>** button shifts the current variable to be the last variable.

Hide
> **Hide** hides the current variable.
> The variable still exists; the editor just stops displaying it.

Delete...
> **Delete...** brings up a dialog that allows you to:
> - Delete the current variable
> - Delete the current observation
> - Delete all observations throughout the dataset that have the
>   same value for the current variable as the current observation

Changing data

1. *Load the* `auto.dta` *dataset that we provide with Stata:*
   Pull down **File** and choose **Open...**, then select `auto.dta` from the `c:\stata` directory (or from wherever Stata is installed).

2. *Enter the editor:*
   Click on the **Data Editor** button or type `edit` in the Command window.

3. *The* **Sort** *button sorts the observations in ascending order of the current variable.*
   Suppose that we want to look at the lowest and highest priced cars. Click once anywhere on the `price` column, and then click on **Sort**. Scroll down to the bottom of the dataset to see the highest priced cars.

4. *The shift button* << (>>) *shifts the current variable to be the first variable (last variable).*
   Let's find the lightest and heaviest cars. Unless you have a big monitor, you will not be able to view both `make` and `weight` together. So, scroll right, click on the `weight` column, and then click on <<. Now, scroll left, select `weight`, and click on **Sort**. Select `make` and click on <<.

```
┌─ Stata Editor ──────────────────────────────────────────── [X] ─┐
│ [Preserve] [Restore]   [ Sort ]  [ << ]  [ >> ]  [ Hide ]  [Delete...]     ▲ │
│                    make[1] = ▐Honda Civic▌                                    │
├──────────┬─────────────┬──────────┬──────────┬──────────┬────┤
│          │    make     │  weight  │  price   │   mpg    │    │
├──────────┼─────────────┼──────────┼──────────┼──────────┼────┤
│   1      │ Honda Civic │  1,760   │  4,499   │    28    │    │
│   2      │ Ford Fiesta │  1,800   │  4,389   │    28    │    │
│   3      │ Plym. Champ │  1,800   │  4,425   │    34    │    │
│   4      │ Renault Le Car │ 1,830 │  3,895   │    26    │    │
│   5      │ VW Rabbit   │  1,930   │  4,697   │    25    │    │
│   6      │ Mazda GLC   │  1,980   │  3,995   │    30    │    │
│   7      │ VW Scirocco │  1,990   │  6,850   │    25    │    │
│   8      │ Datsun 210  │  2,020   │  4,589   │    35    │    │
│   9      │ VW Diesel   │  2,040   │  5,397   │    41    │    │
│  10      │ Subaru      │  2,050   │  3,798   │    35    │    │
│  11      │ Audi Fox    │  2,070   │  6,295   │    23    │    │
│  12      │ Chev. Chevette │ 2,110 │  3,299   │    29    │    │
│  13      │ Dodge Colt  │  2,120   │  3,984   │    30    │    │
│  14      │ Fiat Strada │  2,130   │  4,296   │    21    │    │
│  15      │ VW Dasher   │  2,160   │  7,140   │    23    │    │
│  16      │ Plym. Horizon │ 2,200  │  4,482   │    25    │  ▼ │
└──────────┴─────────────┴──────────┴──────────┴──────────┴────┘
```

Changing data, continued

5. *The* **Hide** *button hides the current variable.*
   The effect of **Hide** is only cosmetic. The variable still exists; it is simply not displayed. Click on the `rep78` column and click on **Hide**.

6. *To change a value, click on the cell, enter the new value, and press Enter or Tab.*
   Try doing this.

7. **Delete...** *allows you to delete one variable, one observation, or all observations that have a certain value.* Select the `trunk` variable. Press **Delete...** and choose **OK**. We dropped `trunk` from the dataset. This change is real; you cannot get `trunk` back unless you cancel all your changes. (But remember, no changes are permanent until you save them on disk. You work with a copy of the data in memory, not with the data file itself.)

8. *Choose* **Restore** *to restore the backup copy of your data.*
   Click on **Restore**. The dataset is now exactly as it was when you began. A backup copy of your dataset is automatically made when you enter the editor. (This can be changed; see [R] **edit** in the *Base Reference Manual*.)

9. **Preserve** *updates the backup copy.*
   Sort again by `mpg`. Click on one of the cells that has `mpg` equal to 14. Press **Delete...**, select the third choice "Delete all 6 obs. where mpg==14", and click on **OK**.

   Click on **Preserve**. Now, make some other changes to the dataset. Click on **Restore**. The changes that you made after the **Preserve** are reversed. There is no way to cancel the changes you made before you pressed **Preserve** (except to reload the original data file).

10. *When you exit the editor, a dialog box will ask you to confirm your changes.*
    If you **Cancel** your changes, the editor reloads the backup copy.

Changing data, continued

11. *You will find output in the Results window documenting the changes you made.*
   What you see are Stata commands that are equivalent to what you did in the editor. The dash in front of the command indicates that the change was done in the editor.

```
. use http://www.stata-press.com/data/r8/auto
(1978 Automobile Data)
. edit
- preserve
- sort price
- sort weight
- drop trunk
- restore
- sort mpg
- drop if mpg == 14
- preserve
- replace weight = 10 in 5
- replace mpg = 10 in 10
- replace rep78 = . in 25
- drop in 3
- restore
```

edit with in and if

Here is one example of using **edit** with selected variables and restricted observations. Chapter 11 contains many examples of using a command with a variable list and **in** and **if**.

```
┌─ Stata Command ─────────────────────────────────────────────────[X]─┐
│                                                                    ▲ │
│ edit make rep78 if rep78 >= .                                        │
│                                                                    ▼ │
└──────────────────────────────────────────────────────────────────────┘
```

```
┌─ Stata Editor ──────────────────────────────────────────────────[X]─┐
│  Preserve  Restore    Sort      <<        >>      Hide     Delete...  ▲ │
│                         make[18] = Buick Opel                          │
│ ┌────────────────────────┬──────────────┬──────────────┬─────────────┐ │
│ │           make         │    rep78      │              │             │ │
│ ├──┬─────────────────────┼──────────────┼──────────────┼─────────────┤ │
│ │18│ Buick Opel          │       .      │              │             │ │
│ │23│ Plym. Sapporo       │       .      │              │             │ │
│ │25│ AMC Spirit          │       .      │              │             │ │
│ │51│ Peugeot 604         │       .      │              │             │ │
│ │52│ Pont. Phoenix       │       .      │              │             │ │
│ └──┴─────────────────────┴──────────────┴──────────────┴─────────────┘ ▼ │
│ ◄                                                                   ► │
└──────────────────────────────────────────────────────────────────────┘
```

Notes

1. You cannot use the **Sort** button if you entered the editor with an **in** or **if** restriction.

2. Stata has multiple missing value indicators: '.' is Stata's default or system missing value indicator, and .a, .b, ..., .z are Stata's extended missing values. Some people use extended missing values to indicate why a certain value is unknown. Other people have no use for extended missing values and just use '.'.

   Stata views missing values as being larger than all numeric nonmissing values, and . < .a < .b < ... < .z; see [U] **15.2.1 Missing values** for full details.

Advice

1. *People who care about data integrity know editors are dangerous — it is easy to accidentally make changes.* Never use **edit** when you just want to look at your data. Use **browse**.

2. *Protect yourself when you* **edit** *data by limiting the dataset's exposure.*
   If you need to change **rep78** only if it is missing, and need to see **make** to make the change, do not press the **Data Editor** button, but type

   ```
   edit make rep78 if rep78>=.
   ```

   It is now impossible for you to change (damage) variables other than **make** and **rep78** and observations other than those with **rep78** equal to missing.

3. *All of this said, Stata's editor is safer than most because it records changes in the Results window.* Use this feature to log your output and make a permanent record of the changes. Then, you can verify that the changes you made are the changes you wanted to make. See Chapter 17 for information on creating log files.

browse

1. *When you want to look at your dataset, but do not want to change it, you can enter the editor in* browse *mode:* Either click on the **Data Browser** button or type browse in the Command window.

2. *In* browse *mode, it is impossible to alter your data.*

3. *You can select variables and restrict observations with* browse, *just like* edit.

	Stata Command	☒
browse make mpg price if foreign == 1		

Stata Browser   ☒

Preserve	Restore	Sort	<<	>>	Hide	Delete...

make[1] = Honda Civic

	make	mpg	price	
1	Honda Civic	28	4,499	
4	Renault Le Car	26	3,895	
5	VW Rabbit	25	4,697	
6	Mazda GLC	30	3,995	
7	VW Scirocco	25	6,850	
8	Datsun 210	35	4,589	
9	VW Diesel	41	5,397	
10	Subaru	35	3,798	
11	Audi Fox	23	6,295	

4. *You can use the* **Sort**, *shift (*<< *and* >>*), and* **Hide** *buttons in* browse *mode.*
   However, just like edit, you cannot sort if you entered browse with an in or if restriction.

5. browse *can be used to do much of what the* list *command does.* But, because you can scroll in the Data Editor, the editor is more convenient for viewing many variables and for scrolling up and down through the observations.

   The list command, which produces output in the Results window, is described in Chapter 11. list is useful for producing listings for your log file (see Chapter 17) and for viewing a few variables quickly in the Results window.

10 Shortcuts: The Review and Variables windows

Entering commands quickly

In this chapter, you will learn:

The Review window shows your past commands.
Click once on a past command and it is copied to the Command window.
Double-click and the command is copied and executed.
To see commands further back, resize the Review window and make it longer.
You can scroll this window.
You can save the contents of this window to a file.

The Variables window shows the current variables.
Click once on a variable and its name is copied to the current target.
(If you double-click, the variable will be copied twice.)
By default, the Command window is the target, but, if a command
dialog is active, one of its fields may be the target.
You can scroll this window.

The Command window follows standard Windows editing style.

Keys for editing in the Command window:

Delete	Deletes the character to the right of the cursor
Backspace	Deletes the character to the left of the cursor
Esc	Deletes the entire command line
Home	Moves the cursor to the beginning of the line
End	Moves the cursor to the end of the line
Page Up	Steps backwards through past command lines
Page Down	Steps forward through past command lines
Tab	Auto-completes a partially typed variable name

(The *Page Up* and *Page Down* keys are equivalent to clicking on
past commands in the Review window.)

You can copy from the Viewer and the Results window to the
clipboard and paste into the Command window.
If the clipboard contains more than one line, only
the first line is pasted into the Command window.

You can also copy from the Graph window to the clipboard.
  • See Chapter 16

The Do-file Editor allows you to submit several commands at once to Stata.
  • See Chapter 15

Continued

More shortcuts

Many Stata commands can be abbreviated.
    Underlining in the syntax diagrams in the command help files indicates the shortest allowable abbreviation.

    Underlining in the syntax diagrams in the *Reference* manuals also indicates the shortest allowable abbreviation.

    Variable names can also be abbreviated in many cases.
- See [U] **14.2 Abbreviation rules** in the *User's Guide* for rules and full details (this manual also gives many examples)

The *F*-keys can be defined to issue commands or to give a list of variable names.
    For example, the *F3* key comes defined as `describe` *Enter*.
    You can easily redefine the *F*-keys to make your own customized shortcuts.

You can resize and rearrange any of Stata's windows.
    You can also save your arrangement or return to the default.
- See Chapter 18

You can close the Review and Variables windows,
    but you cannot close the Command and Results windows.
- See Chapter 18 for details about opening windows, etc.

You can change the fonts in most of the windows:
- See Chapter 18

Standard keyboard shortcuts for menus:
- Press *Alt* and the underlined letter of the menu name
- E.g., for **F**ile, press *Alt-F*

Entering commands quickly

1. *Let's say that you have issued many different commands in a Stata session.*
   Suppose that one of these commands was a `regress` command, and now you want to add another variable to the regression and rerun it.

2. *Use the mouse to move the pointer into the Review window and over the `regress` command.*
   Click once. The previous command will be copied into the Command window.

3. *Move the pointer into the Variables window and click once on `foreign`.*
   The Command window now contains the command you want. Press *Enter* to issue the command.

Another way to quickly enter a variable name is to take advantage of Stata's variable name completion feature. Simply type the first few letters of the variable name in the Command window and press the *Tab* key. Stata will automatically type the rest of the variable name for you. If more than one variable name matches the letters you have typed, Stata will complete as much as it can and beep at you to let you know that you have typed a nonunique variable abbreviation.

Entering commands quickly, continued

4. *If you want to reissue a previous command exactly as typed:*
Just double-click on the command in the Review window. Double-clicking instantly executes the command. It is equivalent to copying the command into the Command window and pressing *Enter*.

5. *You can also use the Review window to quickly fix typos.*
If you make a typo and issue the command without realizing it, you do not have to retype the whole command line again. Just go to the Review window, click once on the errant command to copy it into the Command window, fix the typo, and hit *Enter*.

6. *You can save the Review window contents to a file.*
You can right-click on the Review window and choose **Save Review Contents...** to save your previously issued commands to a file; see [GSW] **D.3 Saving the contents of the Review window as a do-file** for more details.

F-keys

1. *Some of the F-keys are defined to have special meanings.*
For example, pressing *F3* is equivalent to typing `describe` and pressing *Enter*. Pressing *F7* is the same as typing `save` with a space after it — there is no *Enter*, so you can now add the name of the file.

2. *Note: The default definitions of these keys are the same for all versions of Stata — Windows, Macintosh, and Unix.* For Stata for Windows users, the definitions are not particularly useful. However, this really does not matter, because the keys only take on great utility when you define them based on the problem at hand. To see the default definitions, type `macro list`.

3. *How is it that pressing F3 yields the same result as typing* `describe`*?*
When Stata comes up, it silently issues the command

```
global F3 "describe;"
```

to itself. In this context, a semicolon means *Enter*. This makes the *F3* key equivalent to typing `describe` and then pressing *Enter*.

4. *To define the F-keys for yourself, you also use the* `global` *command.*
For example, let's say you plan to run many `logistic` commands, and you don't want to have to type all those letters in `logistic` each time. If you type

```
global F8 "logistic "
```

then hitting *F8* is equivalent to typing `logistic` and a space.

5. *Suppose* that all the logistic regressions that you plan to run will have the dependent variable `outcome` and will always include the three independent variables `drug`, `sex`, and `age`. You want to play around with adding other variables to the regression. You can type

```
global F9 "logistic outcome drug sex age "
```

Now, each time you run a regression, you will only have to press *F9*, then go to the Variables window, click on the names of any additional independent variables, and press *Enter*.

6. *See* [U] **13 Keyboard use** *for more information.*

11 Listing data

list

In this chapter, you will learn:

---

list is similar to browse.
    To list data in the Results window, type: `list`

When a —more— appears in the Results window:
    (as happens when listing a long list)

    To see the next line: Press *Enter*.

    To see the next screen: Press any key, such as *Space Bar*.
    or: Click on the **More** button.
    or: Click on the —more— at the
        bottom of the Results window.

To interrupt a Stata command at any time
and return to the state before you
issued the command: Click on the **Break** button.
    or: Press *Ctrl-Break*.

To list a single variable: `list varname`
    Example: `list displacement`
    You may abbreviate `list`:   `l displacement`
    You may abbreviate *varname*:   `list displ`
    You may use a ~ abbreviation: `list displa~t`

To list any number of variables: `list varname(s)`
    Example: `list make mpg`
    You may abbreviate: `l mak mpg`

To list $varname_i$ through $varname_j$:   `list` $varname_i$`-`$varname_j$
    Example: `list make-mpg`
    You may abbreviate: `l mak-mpg`

To list all variables starting with `pop`:   `list pop*`

To list all variables starting with `ma`,
    a character, and ending in `e`:   `list ma?e`

You may combine any of the above: `list mak-mpg displa~t pop*`

Continued

list, continued

To list the third observation:	`list in 3`
the second from last:	`list in -2`
the last:	`list in -1`
or:	`list in l` (i.e., the letter ℓ)
To list observations 1 through 3:	`list in 1/3`
5 to 17:	`list in 5/17`
3 to third from last:	`list in 3/-3`
You may combine all the above:	`list mpg pop* in 3/-3`
To conditionally list observations:	`list if exp`
Example:	`list if mpg>20`
You may combine all the above:	`list mpg wei pop* if mpg>20`
	`l mpg wei pop* if mpg>20 in 3/-3`
Output from `list` can be logged (as can all output that appears in the Results window):	See Chapter 17.
To specify a minimum abbreviation of 10 characters for variable names:	`list, abbreviate(10)`
To specify divider lines between columns in your listing:	`list, divider`
To specify horizontal separator lines be drawn after every 2 observations:	`list, separator(2)`
To prevent horizontal separator lines from being drawn:	`list, separator(0)`
To draw a horizontal separator line when variable `foreign` changes values:	`list, sepby(foreign)`

list with variable list

```
. list

                   make    price    mpg   weight   gear_r~o    foreign

  1.         VW Rabbit     4697     25     1930       3.78      foreign
  2.           Olds 98     8814     21     4060       2.41     domestic
  3.       Chev. Monza     3667      .     2750       2.73     domestic
  4.                       4099     22     2930       3.58     domestic
  5.        Datsun 510     5079     24     2280       3.54      foreign

  6.       Buick Regal     5189     20     3280       2.93     domestic
  7.        Datsun 810     8129      .     2750       3.55      foreign

. list make mpg price

                   make    mpg    price

  1.         VW Rabbit     25     4697
  2.           Olds 98     21     8814
  3.       Chev. Monza      .     3667
  4.                       22     4099
  5.        Datsun 510     24     5079

  6.       Buick Regal     20     5189
  7.        Datsun 810      .     8129

. list m*

                   make    mpg

  1.         VW Rabbit     25
  2.           Olds 98     21
  3.       Chev. Monza      .
  4.                       22
  5.        Datsun 510     24

  6.       Buick Regal     20
  7.        Datsun 810      .

. list price-weight

          price    mpg   weight

  1.       4697     25     1930
  2.       8814     21     4060
  3.       3667      .     2750
  4.       4099     22     2930
  5.       5079     24     2280

  6.       5189     20     3280
  7.       8129      .     2750
```

list with variable list, continued

```
. list ma?e

                    make

     1.        VW Rabbit
     2.          Olds 98
     3.      Chev. Monza
     4.
     5.        Datsun 510

     6.      Buick Regal
     7.      Datsun 810

. list gear_r~o

             gear_r~o

     1.          3.78
     2.          2.41
     3.          2.73
     4.          3.58
     5.          3.54

     6.          2.93
     7.          3.55
```

Notes

1. list without arguments lists the entire dataset. You can always press the **Break** button to abort a listing.

2. You can list a subset of variables. You can be specific, as in list make mpg price. You can use shorthand: list m* means list all variables starting with m. list price-weight means list variables price through weight in dataset order. list ma?e means list all variables starting with ma, any character, and ending in e.

3. You can list a variable using an abbreviation unique to that variable, as in list gear_r~o. If there is more than one variable that the abbreviation could logically represent, an error message will be returned stating that the abbreviation is ambiguous.

4. You can abbreviate list as l (the letter ℓ).

list with in

```
. list

                make     price     mpg   weight   gear_r~o      foreign

  1.        VW Rabbit      4697      25     1930       3.78      foreign
  2.          Olds 98      8814      21     4060       2.41     domestic
  3.      Chev. Monza      3667       .     2750       2.73     domestic
  4.                       4099      22     2930       3.58     domestic
  5.        Datsun 510     5079      24     2280       3.54      foreign

  6.      Buick Regal      5189      20     3280       2.93     domestic
  7.      Datsun 810       8129       .     2750       3.55      foreign

. list in 1

                make     price     mpg   weight   gear_r~o      foreign

  1.        VW Rabbit      4697      25     1930       3.78      foreign

. list in -1

                make     price     mpg   weight   gear_r~o      foreign

  7.        Datsun 810     8129       .     2750       3.55      foreign

. list in 2/4

                make     price     mpg   weight   gear_r~o      foreign

  2.          Olds 98      8814      21     4060       2.41     domestic
  3.      Chev. Monza      3667       .     2750       2.73     domestic
  4.                       4099      22     2930       3.58     domestic

. list make mpg in -3/-2

                make      mpg

  5.        Datsun 510     24
  6.       Buick Regal     20
```

Notes

1. in restricts the list to a range of observations.

2. Positive numbers count from the top of the dataset. Negative numbers count from the end of the dataset.

3. You may specify both a variable list and an observation range.

list with if

```
. list

                  make     price    mpg    weight    gear_r~o    foreign

  1.          VW Rabbit     4697     25      1930        3.78      foreign
  2.             Olds 98    8814     21      4060        2.41     domestic
  3.         Chev. Monza    3667      .      2750        2.73     domestic
  4.                        4099     22      2930        3.58     domestic
  5.          Datsun 510    5079     24      2280        3.54      foreign

  6.         Buick Regal    5189     20      3280        2.93     domestic
  7.          Datsun 810    8129      .      2750        3.55      foreign

. list if mpg>22

                  make     price    mpg    weight    gear_r~o    foreign

  1.          VW Rabbit     4697     25      1930        3.78      foreign
  3.         Chev. Monza    3667      .      2750        2.73     domestic
  5.          Datsun 510    5079     24      2280        3.54      foreign
  7.          Datsun 810    8129      .      2750        3.55      foreign

. list if mpg>22 & mpg<.

                  make     price    mpg    weight    gear_r~o    foreign

  1.          VW Rabbit     4697     25      1930        3.78      foreign
  5.          Datsun 510    5079     24      2280        3.54      foreign

. list make mpg if mpg>22 | (price>8000 & gear_ratio>3.5)

                  make     mpg

  1.          VW Rabbit     25
  3.         Chev. Monza     .
  5.          Datsun 510    24
  7.          Datsun 810     .

. list make mpg if mpg>22 | (price>8000 & gear_ratio>3.5) in 1/4

                  make     mpg

  1.          VW Rabbit     25
  3.         Chev. Monza     .
```

Notes

1. if expressions may be arbitrarily complicated. & means "and"; | means "or".

2. if may be combined with in (or in with if; the order does not matter).

list with if, continued

3. Stata supports different types of numeric missing values that can be used to specify different reasons a value is unknown. Numeric missing values are represented in Stata by "large positive values". Thus, `mpg>22` evaluates to true when `mpg` equals missing value. You can exclude observations with missing values for `mpg` by adding '`& mpg<.`' to your `if` expression or by using the `missing()` function to test for missing values of `mpg`: '`!missing(mpg)`'. The '`<`' is the symbol for "less than", '`.`' is the missing-value symbol, and '`!`' means "not".

4. The logical operators are

`<`	less than
`<=`	less than or equal
`==`	equal
`>`	greater than
`>=`	greater than or equal
`!=`	not equal (`~=` can also be used)
`&`	and
`|`	or
`!`	not (logical negation; `~` can also be used)
`()`	parentheses specify order of evaluation

By default, `&` is evaluated before `|`; therefore, $a|b\&c$ means $a|(b\&c)$, which is true if a is true or if b and c are both true. To specify a or b being true, and c being true, too, type $(a|b)\&c$.

list with if, common mistakes

```
. list

            make    price    mpg    weight    gear_r~o    foreign

  1.     VW Rabbit     4697     25      1930        3.78     foreign
  2.        Olds 98    8814     21      4060        2.41    domestic
  3.    Chev. Monza    3667      .      2750        2.73    domestic
  4.                   4099     22      2930        3.58    domestic
  5.     Datsun 510    5079     24      2280        3.54     foreign

  6.    Buick Regal    5189     20      3280        2.93    domestic
  7.     Datsun 810    8129      .      2750        3.55     foreign

. list if mpg=21
=exp not allowed
r(101);

. list if mpg==21

            make    price    mpg    weight    gear_r~o    foreign

  2.       Olds 98     8814     21      4060        2.41    domestic

. list if mpg==21 if weight>4000
invalid syntax
r(198);

. list if mpg==21 and weight>4000
invalid 'and'
r(198);

. list if mpg==21 & weight>4000

            make    price    mpg    weight    gear_r~o    foreign

  2.       Olds 98     8814     21      4060        2.41    domestic
```

Notes

1. Tests of equality are specified with double equal signs, not single — if mpg==21, not if mpg=21. Single equal signs, you will learn, are used for assignment.

2. Joint tests are specified with &, not multiple ifs — if mpg==21 & weight>4000, not if mpg==21 if weight>4000.

3. Use & and |, not the words and and or.

list with if, common mistakes, continued

```
. list

                  make     price    mpg   weight   gear_r~o    foreign

     1.       VW Rabbit      4697     25     1930       3.78     foreign
     2.         Olds 98      8814     21     4060       2.41    domestic
     3.     Chev. Monza      3667      .     2760       2.73    domestic
     4.                      4099     22     2930       3.58    domestic
     5.      Datsun 510      5079     24     2280       3.54     foreign

     6.     Buick Regal      5189     20     3280       2.93    domestic
     7.      Datsun 810      8129      .     2750       3.55     foreign

. list if make==Datsun 510
Datsun not found
r(111);

. list if make=="Datsun 510"

                  make     price    mpg   weight   gear_r~o    foreign

     5.      Datsun 510      5079     24     2280       3.54     foreign
```

Note

Tests with strings are allowed, but if you specify the contents of the string, the contents are enclosed in double quotes: if make=="Datsun 510".

list with if, common mistakes, continued

```
. list if foreign==domestic
domestic not found
r(111);
. list if foreign==0
```

	make	price	mpg	weight	gear_r~o	foreign
2.	Olds 98	8814	21	4060	2.41	domestic
3.	Chev. Monza	3667	.	2750	2.73	domestic
4.		4099	22	2930	3.58	domestic
6.	Buick Regal	5189	20	3280	2.93	domestic

```
. list if foreign==0, nolabel
```

	make	price	mpg	weight	gear_r~o	foreign
2.	Olds 98	8814	21	4060	2.41	0
3.	Chev. Monza	3667	.	2750	2.73	0
4.		4099	22	2930	3.58	0
6.	Buick Regal	5189	20	3280	2.93	0

Notes

1. Watch out for value labels — they look like strings but are not. Variable `foreign` takes on values 0 and 1, and we have merely labeled 0 "domestic" and 1 "foreign" (see Chapter 8).

2. To see the underlying numeric values of variables with labeled values, use the `nolabel` option on the `list` or `edit` commands (see [R] **list** and [R] **edit**). You can also determine the correspondence between labels and numeric values with the `label list` command (see [R] **label**).

Controlling the list output

```
. sort foreign
. list, sepby(foreign)
```

	make	price	mpg	weight	gear_r~o	foreign
1.		4099	22	2930	3.58	domestic
2.	Olds 98	8814	21	4060	2.41	domestic
3.	Buick Regal	5189	20	3280	2.93	domestic
4.	Chev. Monza	3667	.	2750	2.73	domestic
5.	Datsun 510	5079	24	2280	3.54	foreign
6.	Datsun 810	8129	.	2750	3.55	foreign
7.	VW Rabbit	4697	25	1930	3.78	foreign

```
. list make weight gear, abbreviate(10)
```

	make	weight	gear_ratio
1.		2930	3.58
2.	Olds 98	4060	2.41
3.	Buick Regal	3280	2.93
4.	Chev. Monza	2750	2.73
5.	Datsun 510	2280	3.54
6.	Datsun 810	2750	3.55
7.	VW Rabbit	1930	3.78

```
. list, divider
```

	make	price	mpg	weight	gear_r~o	foreign
1.		4099	22	2930	3.58	domestic
2.	Olds 98	8814	21	4060	2.41	domestic
3.	Buick Regal	5189	20	3280	2.93	domestic
4.	Chev. Monza	3667	.	2750	2.73	domestic
5.	Datsun 510	5079	24	2280	3.54	foreign
6.	Datsun 810	8129	.	2750	3.55	foreign
7.	VW Rabbit	4697	25	1930	3.78	foreign

Controlling the list output, continued

```
. list, separator(3)

                make    price    mpg    weight    gear_r~o    foreign

  1.                     4099     22      2930       3.58     domestic
  2.         Olds 98     8814     21      4060       2.41     domestic
  3.     Buick Regal     5189     20      3280       2.93     domestic

  4.     Chev. Monza     3667      .      2750       2.73     domestic
  5.     Datsun 510      5079     24      2280       3.54      foreign
  6.     Datsun 810      8129      .      2750       3.55      foreign

  7.       VW Rabbit     4697     25      1930       3.78      foreign
```

More

When you see a —more— at the bottom of the Results window, it means that there is more information to be displayed. This happens, for example, when listing a large number of observations.

```
. list make mpg

          make                mpg

    1.    Linc. Continental    12
    2.    Linc. Mark V         12
    3.    Linc. Versailles     14
    4.    Merc. XR-7           14
    5.    Cad. Deville         14

    6.    Peugeot 604          14
    7.    Cad. Eldorado        14
    8.    Merc. Cougar         14
    9.    Buick Electra        15
   10.    Merc. Marquis        15

   11.    Olds Toronado        16
   12.    Buick Riviera        16
   13.    Dodge Magnum         16
   14.    Chev. Impala         16
   15.    Volvo 260            17

   16.    Audi 5000            17
   17.    AMC Pacer            17
   18.    Dodge St. Regis      17
   19.    Pont. Firebird       18
   20.    Olds Delta 88        18

  —more—
```

If you want to see the next screen of text, press any key, such as the *Space Bar*, click on the **More** button, or click on the blue —more— at the bottom of the Results window. To see just the next line of text, press *Enter*.

Break

If you want to interrupt a Stata command, click on the **Break** button, or press *Ctrl-Break*.

```
. list make mpg

       | make                 mpg |
     1.| Linc. Continental     12 |
     2.| Linc. Mark V          12 |
     3.| Linc. Versailles      14 |
     4.| Merc. XR-7            14 |
     5.| Cad. Deville         14 |
       |                          |
     6.| Peugeot 604          14 |
     7.| Cad. Eldorado        14 |
     8.| Merc. Cougar         14 |
     9.| Buick Electra        15 |
    10.| Merc. Marquis        15 |
       |                          |
    11.| Olds Toronado        16 |
    12.| Buick Riviera        16 |
    13.| Dodge Magnum         16 |
    14.| Chev. Impala         16 |
    15.| Volvo 260            17 |
       |                          |
    16.| Audi 5000            17 |
    17.| AMC Pacer            17 |
    18.| Dodge St. Regis      17 |
    19.| Pont. Firebird       18 |
    20.| Olds Delta 88        18 |

—Break—
r(1);
```

Note

It is always safe to click **Break**. After you click **Break**, the state of the system is the same as if you had never issued the command.

12 Creating new variables

generate and replace

In this chapter, you will learn:

To create a new variable that is an algebraic expression of other variables:	generate *newvar* = *exp*
generate may be abbreviated:	g *newvar* = *exp*
To change the contents of an existing variable:	replace *oldvar* = *exp*
replace may not be abbreviated.	

exp is an algebraic expression that is a combination of existing variables, operators, and functions.

Operators:

	Arithmetic		Logical		Relational (numeric and string)
+	addition	!	not	>	greater than
−	subtraction	|	or	<	less than
*	multiplication	&	and	>=	> or equal
/	division			<=	< or equal
^	power			==	equal
				!=	not equal
+	string concatenation				

Stata has many mathematical, statistical, string, date, time-series, and programming functions. See [R] **functions** for a complete list and full details.

generate

```
. list price mpg

        price    mpg

  1.     4697     25
  2.     8814     21
  3.     3667      .
  4.     4099     22
  5.     5079     24

  6.     5189     20
  7.     8129      .

. gen logpr = ln(price)
. gen ratio = price/mpg
(2 missing values generated)
. gen silly = ((price+100)/ln(mpg-3))^2
(2 missing values generated)
. list price mpg logpr ratio silly

        price    mpg      logpr      ratio      silly

  1.     4697     25   8.454679     187.88    2408405
  2.     8814     21   9.084097   419.7143    9511256
  3.     3667      .   8.207129          .          .
  4.     4099     22   8.318499   186.3182    2033699
  5.     5079     24   8.532869    211.625    2893700

  6.     5189     20   8.554296     259.45    3484886
  7.     8129      .   9.003193          .          .
```

Notes

1. The form of the `generate` command is `generate` *newvar* = *exp*, where *newvar* is a new variable name (it cannot be the name of a variable that already exists) and *exp* is any valid expression.

2. The `generate` command may be abbreviated as `g`, `ge`, `gen`, etc.

3. An expression is a combination of existing variables, operators, and functions. Expressions can be made as complicated as you want.

4. Calculation on a missing value yields a missing value and so does division by zero, etc.

5. If missing values are generated, the number of missing values in *newvar* is always reported. The lack of a mention means no missing values resulted.

replace

```
. list make weight price foreign, nolabel

              make    weight    price    foreign

  1.      VW Rabbit      1930     4697          1
  2.         Olds 98     4060     8814          0
  3.    Chev. Monza      2750     3667          0
  4.                     2930     4099          0
  5.      Datsun 510     2280     5079          1

  6.     Buick Regal     3280     5189          0
  7.     Datsun 810      2750     8129          1

. gen weight = weight/1000
weight already defined
r(110);

. replace weight = weight/1000
(7 real changes made)
```

Notes

1. Use `generate` to create new variables and use `replace` to change the contents of existing variables. Stata requires this so that you do not accidentally modify your data.

2. The `replace` command cannot be abbreviated. Stata generally requires you to spell out completely any command that can alter your existing data.

replace, continued

You want to create a new variable `predprice`, which will be the predicted price of the cars in the following year. You estimate that domestic cars will increase in price by 5% and foreign cars by 10%. First, use `generate` to compute the predicted domestic car prices. Then, use `replace` to change the missing values for the foreign cars to their proper values.

```
. gen predprice = 1.05*price if foreign == 0
(3 missing values generated)
. replace predprice = 1.1*price if foreign == 1
(3 real changes made)
. list make weight price predprice foreign, nolabel
```

	make	weight	price	predpr~e	foreign
1.	VW Rabbit	1.93	4697	5166.7	1
2.	Olds 98	4.06	8814	9254.7	0
3.	Chev. Monza	2.75	3667	3850.35	0
4.		2.93	4099	4303.95	0
5.	Datsun 510	2.28	5079	5586.9	1
6.	Buick Regal	3.28	5189	5448.45	0
7.	Datsun 810	2.75	8129	8941.9	1

generate with string variables

```
. list make foreign, nolabel

              make    foreign

  1.      VW Rabbit          1
  2.        Olds 98          0
  3.    Chev. Monza          0
  4.                         0
  5.     Datsun 510          1

  6.    Buick Regal          0
  7.     Datsun 810          1

. gen origin = "D" if foreign == 0
(3 missing values generated)
. replace origin = "F" if foreign == 1
(3 real changes made)
. list make foreign origin

              make     foreign    origin

  1.      VW Rabbit    foreign        F
  2.        Olds 98   domestic        D
  3.    Chev. Monza   domestic        D
  4.                  domestic        D
  5.     Datsun 510    foreign        F

  6.    Buick Regal   domestic        D
  7.     Datsun 810    foreign        F

. describe origin

                 storage   display     value
variable name     type     format      label      variable label
-----------------------------------------------------------------------
origin            str1      %9s
```

Note

1. Stata is smart. When you generate a variable and the expression evaluates to a string, Stata creates a string variable with a storage type as long as necessary, and no longer than that. origin is a str1.

generate with string variables, continued

```
. gen makeorigin = make + " " + origin
. gen word2 = substr(make, index(make," ")+1, .)
(1 missing value generated)
. list make origin makeorigin word2
```

	make	origin	makeorigin	word2
1.	VW Rabbit	F	VW Rabbit F	Rabbit
2.	Olds 98	D	Olds 98 D	98
3.	Chev. Monza	D	Chev. Monza D	Monza
4.		D	D	
5.	Datsun 510	F	Datsun 510 F	510
6.	Buick Regal	D	Buick Regal D	Regal
7.	Datsun 810	F	Datsun 810 F	810

Notes

1. The operator '+' when applied to string variables will concatenate the strings (i.e., join them together). The expression "this" + "that" results in the string "thisthat". When the variable makeorigin was generated, a space (" ") was added between the two strings.

2. index(s_1,s_2) produces an integer equal to the first position in s_1 in which s_2 is found, or 0 if not found.

   substr(s,c_0,c_1) produces a string equal to the columns c_0 to $c_0 + c_1$ of s. If $c_1 = .$, it gives a string from c_0 to end of string.

   Therefore, substr(s,index(s," ")+1,.) produces s with its first word removed.

13 Deleting variables and observations

clear, drop, and keep

In this chapter, you will learn:

To drop all the data in memory:	`drop _all`
To drop all the data in memory and clear other memory areas used by graphs, dialogs, matrices, etc.:	`clear`
To drop a single variable:	`drop` *varname*
Example:	`drop weight`
Example:	`drop gear_r~o`
To drop any number of variables:	`drop` *varname(s)*
Example:	`drop make mpg`
Example:	`drop ma?e`
To drop *varname$_i$* through *varname$_j$*:	`drop` *varname$_i$*-*varname$_j$*
Example:	`drop make-mpg`
To drop all variables starting with `pop`:	`drop pop*`
You may combine all the above:	`drop make-mpg weight pop*`
To drop the first observation:	`drop in 1`
the second:	`drop in 2`
the third:	`drop in 3`
the second from last:	`drop in -2`
the last:	`drop in -1`
or:	`drop in l` (i.e., the letter ℓ)
To drop observations 1 through 3:	`drop in 1/3`
5 to 17:	`drop in 5/17`
3 to third from last:	`drop in 3/-3`
To conditionally drop observations:	`drop if` *exp*
Example:	`drop if mpg>20`
You may combine `if` and `in`:	`drop if mpg>20 in 3/-3`
`keep` works like `drop`, except you specify the variables or observations to keep:	`keep if mpg>20 in 3/-3`
To make changes permanent, resave the dataset:	Choose **Save** under the **File** menu.
or, alternatively, you can type:	`save` *filename*`, replace`

drop _all and clear

Both `drop _all` and `clear` eliminate the dataset from memory:

```
. drop _all
. list
```

or

```
. clear
. list
```

They differ in that

1. `drop _all` drops just the dataset from memory.

2. `clear` drops the dataset, and it drops all value label definitions, scalars and matrices, constraints and equations, programs, previous estimation results, dialogs, and graphs. `clear`, in effect, resets Stata. The first time you use `clear` while you have a graph or dialog box up, you may be surprised when that graph or dialog box closes; this is necessary so that Stata can free all memory that is being used.

drop

```
. use afewcars
(A few 1978 cars)
. list
```

	make	price	mpg	weight	gear_r~o	foreign
1.	VW Rabbit	4697	25	1930	3.78	foreign
2.	Olds 98	8814	21	4060	2.41	domestic
3.	Chev. Monza	3667	.	2750	2.73	domestic
4.		4099	22	2930	3.58	domestic
5.	Datsun 510	5079	24	2280	3.54	foreign
6.	Buick Regal	5189	20	3280	2.93	domestic
7.	Datsun 810	8129	.	2750	3.55	foreign

```
. drop in 1/3
(3 observations deleted)
. list
```

	make	price	mpg	weight	gear_r~o	foreign
1.		4099	22	2930	3.58	domestic
2.	Datsun 510	5079	24	2280	3.54	foreign
3.	Buick Regal	5189	20	3280	2.93	domestic
4.	Datsun 810	8129	.	2750	3.55	foreign

```
. drop if mpg>21
(3 observations deleted)
. list
```

	make	price	mpg	weight	gear_r~o	foreign
1.	Buick Regal	5189	20	3280	2.93	domestic

```
. drop gear_ratio
. list
```

	make	price	mpg	weight	foreign
1.	Buick Regal	5189	20	3280	domestic

drop, continued

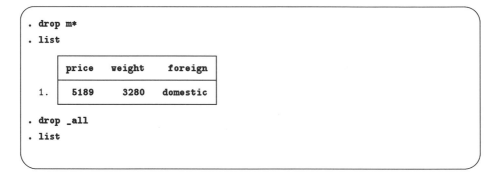

```
. drop m*
. list

        price    weight    foreign

  1.    5189      3280     domestic

. drop _all
. list
```

Note

To make changes permanent, you must resave the dataset. Since we do not in the above example, afewcars.dta remains unchanged.

keep

```
. use afewcars, clear
(A few 1978 cars)
. list
```

	make	price	mpg	weight	gear_r~o	foreign
1.	VW Rabbit	4697	25	1930	3.78	foreign
2.	Olds 98	8814	21	4060	2.41	domestic
3.	Chev. Monza	3667	.	2750	2.73	domestic
4.		4099	22	2930	3.58	domestic
5.	Datsun 510	5079	24	2280	3.54	foreign
6.	Buick Regal	5189	20	3280	2.93	domestic
7.	Datsun 810	8129	.	2750	3.55	foreign

```
. keep in 4/7
(3 observations deleted)
. list
```

	make	price	mpg	weight	gear_r~o	foreign
1.		4099	22	2930	3.58	domestic
2.	Datsun 510	5079	24	2280	3.54	foreign
3.	Buick Regal	5189	20	3280	2.93	domestic
4.	Datsun 810	8129	.	2750	3.55	foreign

```
. keep if mpg<=21
(3 observations deleted)
. list
```

	make	price	mpg	weight	gear_r~o	foreign
1.	Buick Regal	5189	20	3280	2.93	domestic

Note

To make changes permanent, you must resave the dataset. Since we do not in the above example, afewcars.dta remains unchanged.

Notes

14 A sample session

Putting it together

In this chapter, you will learn:

Most of Stata's commands share a common syntax, which is

> [by *varlist*:] *command* [*varlist*] [if *exp*] [in *range*] [, *options*]

where items enclosed in square brackets are optional.

Example: `list` of Chapter 11:

```
list
list mpg weight
list if mpg>20
list mpg weight if mpg>20
list in 1/10
list mpg weight in 1/10
list mpg weight if mpg>20 in 1/10
```

by repeats the command for each set of values of its *varlist*:

```
summarize mpg
by rep78: summarize mpg
```

The first command reports the mean, etc., of `mpg`.
The second command reports the mean, etc., of `mpg` for each value of `rep78`.

Most commands allow `by`, but some do not.

Determine which are which by selecting **Stata command...** from the **Help** menu, entering *commandname*, and examining the syntax diagram in the help file.

See **Help** for `summarize` and discover that `by` is allowed.
See **Help** for `label` and discover that `by` is not allowed.

A single comma sets off a command's options from the rest of the command.

```
list mpg weight, divider separator(3)
```

is the correct way to specify options `divider` and `separator()`.

```
list mpg weight, divider, separator(3)
```

with commas between the options is incorrect.

Warning:

The purpose of this sample session is to show you a little of Stata in action.

Please do not assume that because we do not demonstrate a feature,
Stata cannot do it.
Stata has lots of other commands.

Please do not assume that because we demonstrate a command,
we demonstrate all aspects of that command.
Stata has lots of options.

In all cases, see the *Reference* manuals for a complete description of Stata's
features. When in doubt, consult the indices located at the back of the manuals.

Sample session

In this chapter, we will use the `auto.dta` file which was shipped with Stata. If you wish to follow along, you must load this dataset. Launch Stata and choose **Open** from the **File** menu. Select the `auto.dta` file from the directory in which you installed Stata. Alternatively, you may load the `auto.dta` file directly from the Internet; you will learn more about this in Chapter 20.

```
. use http://www.stata-press.com/data/r8/auto, clear
(1978 Automobile Data)
```

The data that we loaded contain

```
. describe
Contains data from http://www.stata-press.com/data/r8/auto.dta
  obs:            74                          1978 Automobile Data
  vars:           12                          14 Oct 2002 09:02
  size:        3,478 (99.5% of memory free)   (_dta has notes)

              storage   display     value
variable name   type    format      label      variable label

make            str18   %-18s                   Make and Model
price           int     %8.0gc                  Price
mpg             int     %8.0g                    Mileage (mpg)
rep78           int     %8.0g                    Repair Record 1978
headroom        float   %6.1f                    Headroom (in.)
trunk           int     %8.0g                    Trunk space (cu. ft.)
weight          int     %8.0gc                  Weight (lbs.)
length          int     %8.0g                    Length (in.)
turn            int     %8.0g                    Turn Circle (ft.)
displacement    int     %8.0g                    Displacement (cu. in.)
gear_ratio      float   %6.2f                    Gear Ratio
foreign         byte    %8.0g       origin       Car type

Sorted by:  foreign
```

Listing can be informative

Here are some of our data:

```
. list make mpg in 1/10
```

	make	mpg
1.	AMC Concord	22
2.	AMC Pacer	17
3.	AMC Spirit	22
4.	Buick Century	20
5.	Buick Electra	15
6.	Buick LeSabre	18
7.	Buick Opel	26
8.	Buick Regal	20
9.	Buick Riviera	16
10.	Buick Skylark	19

Question: Which cars yield the lowest gas mileage?

```
. sort mpg
. l make mpg in 1/5
```

	make	mpg
1.	Linc. Mark V	12
2.	Linc. Continental	12
3.	Merc. Cougar	14
4.	Merc. XR-7	14
5.	Linc. Versailles	14

Which 5 cars yield the highest gas mileage?

```
. l make mpg in -5/-1
```

	make	mpg
70.	Toyota Corolla	31
71.	Plym. Champ	34
72.	Subaru	35
73.	Datsun 210	35
74.	VW Diesel	41

Descriptive statistics

Question: Not being familiar with 1978 prices, what is the average price of a car in this data?

```
. summarize price
    Variable |      Obs        Mean    Std. Dev.       Min        Max
-------------+--------------------------------------------------------
       price |       74    6165.257    2949.496       3291      15906
```

Aside: `summarize` works like `list` — without arguments it provides a summary of all of the data:

```
. summarize
    Variable |      Obs        Mean    Std. Dev.       Min        Max
-------------+--------------------------------------------------------
        make |        0
       price |       74    6165.257    2949.496       3291      15906
         mpg |       74     21.2973    5.785503         12         41
       rep78 |       69    3.405797    .9899323          1          5
    headroom |       74    2.993243    .8459948        1.5          5
-------------+--------------------------------------------------------
       trunk |       74    13.75676    4.277404          5         23
      weight |       74    3019.459    777.1936       1760       4840
      length |       74    187.9324    22.26634        142        233
        turn |       74    39.64865    4.399354         31         51
displacement |       74    197.2973    91.83722         79        425
-------------+--------------------------------------------------------
  gear_ratio |       74    3.014865    .4562871       2.19       3.89
     foreign |       74    .2972973    .4601885          0          1
```

Note

`make` has 0 observations because it is a string — calculating a mean is undefined but is not an error. `rep78` has only 69 observations because for five cars, it is missing.

Descriptive statistics, continued

Question: What is the average price of cars that are below and above the mean MPG?

```
. summarize price if mpg<21.3
    Variable |       Obs        Mean    Std. Dev.        Min        Max
-------------+--------------------------------------------------------
       price |        43     7091.86    3425.019       3291      15906
. summarize price if mpg>=21.3
    Variable |       Obs        Mean    Std. Dev.        Min        Max
-------------+--------------------------------------------------------
       price |        31    4879.968    1344.659       3299       9735
```

Aside: if can be suffixed to almost all commands. This is one of Stata's more useful features.

Question: What is the median MPG?

```
. summarize mpg, detail

                         Mileage (mpg)
-------------------------------------------------------------
      Percentiles      Smallest
 1%          12              12
 5%          14              12
10%          14              14       Obs                  74
25%          18              14       Sum of Wgt.          74

50%          20                       Mean            21.2973
                         Largest      Std. Dev.      5.785503
75%          25              34
90%          29              35       Variance       33.47205
95%          34              35       Skewness       .9487176
99%          41              41       Kurtosis       3.975005
```

Answer: 20.

Descriptive statistics, continued

Aside: The ', `detail`' at the end of the `summarize` command is called an option. Most Stata commands share a common syntax:

$$[\text{by } varlist:] \; command \; [varlist] \; [\text{if } exp] \; [\text{in } range] \; [, \; options \;]$$

Square brackets mean optional. Thus,

command by itself is valid:	`summarize`
command followed by a *varlist* (variable list) is valid:	`summarize mpg` `summarize mpg weight`
command with `if` (with or without a *varlist*) is valid:	`summarize if mpg>20` `summarize mpg weight if mpg>20`

and so on.

If *varlist* is not specified, all the variables are used.

`if` and `in` restrict the data on which the command is run.

options modify what the command does.

Each command's syntax is found in the *Reference* manuals. You can learn about `summarize` in [R] **summarize**, or choose **Stata command...** from the **Help** menu and enter `summarize`.

Descriptive statistics, continued

Our dataset contains variable `foreign` that is 0 if the car was manufactured in the United States or Canada and 1 otherwise.

Problem: Obtain summary statistics for price and MPG for each value of foreign.

There are two solutions to this problem:

1. Type in the commands:

```
summarize price mpg if foreign==0
summarize price mpg if foreign==1
```

2. Or, you could do the following:

```
. sort foreign
. by foreign: summarize price mpg
```

-> foreign = Domestic

Variable	Obs	Mean	Std. Dev.	Min	Max
price	52	6072.423	3097.104	3291	15906
mpg	52	19.82692	4.743297	12	34

-> foreign = Foreign

Variable	Obs	Mean	Std. Dev.	Min	Max
price	22	6384.682	2621.915	3748	12990
mpg	22	24.77273	6.611187	14	41

Descriptive statistics, continued

Explanation: Stata's general command syntax is

$$[\text{by } \textit{varlist}:] \ \textit{command} \ [\textit{varlist}] \ [\text{if } \textit{exp}] \ [\text{in } \textit{range}] \ [, \ \textit{options} \ ]$$

When by is placed in front of a command, the command is repeated for each set of values of the specified variables.

Most commands allow by but some do not. Online help always begins with a command's syntax diagram. So, you can quickly tell whether by is allowed by clicking on **Help**, selecting **Stata command...**, and entering *commandname*. If by is allowed, you will see

```
by ...: may be used with ...
```

immediately following the syntax diagram.

More explanation: Actually, Stata's command syntax includes even more than we have shown you. See [U] **14 Language syntax**.

Note on by: To use by, the data should be sorted by the by-variables. That is why we typed sort foreign before typing by foreign: .... Additional features of the by prefix, such as sorting, can be found in [R] **by**.

Minor note: How is it that foreign takes on values 0 and 1, and yet on the output, its values appear to be "Foreign" and "Domestic"? foreign has a value label, associating $0 \leftrightarrow$ Domestic and $1 \leftrightarrow$ Foreign; see Chapter 8.

More on by

Problem: It appears that the average MPG of domestic and foreign cars differs. Test the hypothesis that the means are equal.

```
. ttest mpg, by(foreign)
Two-sample t test with equal variances
```

Group	Obs	Mean	Std. Err.	Std. Dev.	[95% Conf. Interval]	
Domestic	52	19.82692	.657777	4.743297	18.50638	21.14747
Foreign	22	24.77273	1.40951	6.611187	21.84149	27.70396
combined	74	21.2973	.6725511	5.785503	19.9569	22.63769
diff		-4.945804	1.362162		-7.661225	-2.230384

```
Degrees of freedom: 72
                 Ho: mean(Domestic) - mean(Foreign) = diff = 0
     Ha: diff < 0               Ha: diff != 0               Ha: diff > 0
       t =  -3.6308               t =  -3.6308               t =  -3.6308
     P < t =   0.0003          P > |t| =   0.0005          P > t =   0.9997
```

Syntax explanation: The by in this command is not the by prefix described on the previous page; it is a by() option.

by as a prefix — if the command allows it — repeats the command on each subsample of data. If there are three groups, the results will be as if we created three separate datasets and applied the command to each.

by() as an option — if the command allows it — informs the command what groups to use, but the command operates on the data *as a whole*. ttest mpg, by(foreign) tests that the means of mpg in the groups defined by foreign are equal.

by ...: ttest, performs separate *t* tests within the groups. So, by rep78: ttest mpg, by(foreign) would report five equality-of-means tests, one for each value of rep78, provided that each group of rep78 contained both foreign cars and domestic cars. Each would be a test of the equality of means for domestic and foreign cars, and each would be calculated independently. For rep78 == 1, there are only two cars, and they are both domestic. If you were to issue the command by rep78: ttest mpg, by(foreign), you would receive an error message indicating that a *t* test is not appropriate.

More on by, continued

Just remember: by in front of a command repeats the command for each group. by() on the right of a command does something else, defined in the command's documentation. The key is an additional line following the syntax diagrams for a command. If the command allows the by prefix, this will be stated both in the online help and in the manuals. The syntax diagram for ttest is

ttest *varname* [if *exp*] [in *range*], by(*groupvar*) [unequal welch level(#)]

followed by

by ...: may be used with ttest (but not with ttesti).

The by prefix is indicated as being allowed — we know what it does. by() is mentioned on the right in the command syntax; we must read the documentation to see what it does.

Another aside: We showed only one of the three syntaxes for ttest. See the help file for ttest or [R] **ttest** for a complete description of ttest.

Analysis note: We have established that in 1978 domestic cars had poorer gas mileage than foreign cars.

Descriptive statistics, making tables

Problem: Obtain counts of the number of domestic and foreign cars.

```
. tabulate foreign
    Car type |     Freq.      Percent         Cum.
-------------+-----------------------------------
    Domestic |        52        70.27        70.27
     Foreign |        22        29.73       100.00
-------------+-----------------------------------
       Total |        74       100.00
```

Problem: The dataset contains variable rep78 recording each car's frequency-of-repair record (1 = poor, ..., 5 = excellent). Obtain frequency counts.

```
. tabulate rep78
     Repair |
Record 1978 |     Freq.      Percent         Cum.
------------+-----------------------------------
          1 |         2         2.90         2.90
          2 |         8        11.59        14.49
          3 |        30        43.48        57.97
          4 |        18        26.09        84.06
          5 |        11        15.94       100.00
------------+-----------------------------------
      Total |        69       100.00
```

Problem: We have 74 cars; only 69 have frequency-of-repair records recorded. List the cars for which data is missing.

```
. list make if rep78>=.

        |          make |
        |---------------|
    14. |  Plym. Sapporo |
    17. |     Buick Opel |
    23. |  Pont. Phoenix |
    27. |     AMC Spirit |
    72. |    Peugeot 604 |
```

Note

list make if rep78>=. is equivalent to list make if missing(rep78); remember that Stata stores missing values as the largest possible numeric values. See [U] **15.2.1 Missing values** for full details.

Descriptive statistics, making tables, continued

Problem: Compare frequency-of-repair records for domestic and foreign cars (i.e., make a two-way table).

```
. tabulate rep78 foreign
  Repair
  Record            Car type
    1978     Domestic    Foreign  |     Total

       1           2          0   |        2
       2           8          0   |        8
       3          27          3   |       30
       4           9          9   |       18
       5           2          9   |       11

   Total          48         21   |       69
```

Problem: Domestic cars appear to have poorer frequency-of-repair records. Is the difference statistically significant? Obtain a χ^2 (even though there are not at least 5 cars expected in each cell):

```
. tabulate rep78 foreign, chi2
  Repair
  Record            Car type
    1978     Domestic    Foreign  |     Total

       1           2          0   |        2
       2           8          0   |        8
       3          27          3   |       30
       4           9          9   |       18
       5           2          9   |       11

   Total          48         21   |       69
        Pearson chi2(4) =  27.2640   Pr = 0.000
```

Analysis note: We find that frequency-of-repair records differ between domestic and foreign cars. In 1978, domestic cars appear to be poorer in this regard.

Descriptive statistics, making tables, continued

Aside: `tabulate` provides options to display any or all of the row percentages, column percentages, and cell percentages along with, or instead of, the frequencies.

The option `nofreq` suppresses the frequencies; option `row` adds row percentages; option `col` adds column percentages; and option `cell` adds cell percentages.

```
. tabulate rep78 foreign, chi2 row col

  ┌─────────────────┐
  │ Key             │
  ├─────────────────┤
  │     frequency   │
  │  row percentage │
  │ column percentage│
  └─────────────────┘

  Repair │
  Record │         Car type
    1978 │  Domestic      Foreign  │      Total
  ───────┼──────────────────────────┼────────────
       1 │        2            0    │         2
         │   100.00         0.00    │    100.00
         │     4.17         0.00    │      2.90
  ───────┼──────────────────────────┼────────────
       2 │        8            0    │         8
         │   100.00         0.00    │    100.00
         │    16.67         0.00    │     11.59
  ───────┼──────────────────────────┼────────────
       3 │       27            3    │        30
         │    90.00        10.00    │    100.00
         │    56.25        14.29    │     43.48
  ───────┼──────────────────────────┼────────────
       4 │        9            9    │        18
         │    50.00        50.00    │    100.00
         │    18.75        42.86    │     26.09
  ───────┼──────────────────────────┼────────────
       5 │        2            9    │        11
         │    18.18        81.82    │    100.00
         │     4.17        42.86    │     15.94
  ───────┼──────────────────────────┼────────────
   Total │       48           21    │        69
         │    69.57        30.43    │    100.00
         │   100.00       100.00    │    100.00

         Pearson chi2(4) =  27.2640   Pr = 0.000
```

Note

We type `tabulate ..., chi2 row col`, not `tabulate ..., chi2, row, col`. One comma sets off the options from the command; commas do not separate options from each other. The order of the options is irrelevant.

Descriptive statistics, correlation matrices

Question: What is the correlation between MPG and weight of car?

```
. correlate mpg weight
(obs=74)
              |     mpg   weight

         mpg |  1.0000
      weight | -0.8072   1.0000
```

Problem: Compare the correlation for domestic and foreign cars.

```
. correlate mpg weight if foreign==0
(obs=52)
              |     mpg   weight

         mpg |  1.0000
      weight | -0.8759   1.0000

. correlate mpg weight if foreign==1
(obs=22)
              |     mpg   weight

         mpg |  1.0000
      weight | -0.6829   1.0000
```

Note

We could have obtained this by typing by foreign: correlate mpg weight instead.

Descriptive statistics, correlation matrices, continued

Aside: We can produce correlation matrices containing as many variables as we wish.

```
. correlate mpg weight price length displacement
(obs=74)
                       mpg    weight     price    length displa~t

         mpg        1.0000
      weight       -0.8072    1.0000
       price       -0.4686    0.5386    1.0000
      length       -0.7958    0.9460    0.4318    1.0000
displacement       -0.7056    0.8949    0.4949    0.8351    1.0000
```

Graphing data

Problem: We know that the average MPG of domestic and foreign cars differs. We have learned that domestic and foreign cars differ in other ways as well, such as in frequency-of-repair record. We found a negative correlation of MPG and weight — as we would expect — but the correlation appears stronger for domestic cars. Examine, with an eye toward modeling, the relationship between MPG and weight. Begin with a graph.

```
. scatter mpg weight
```

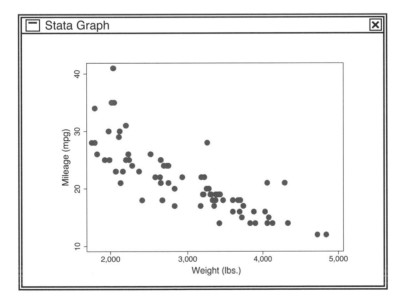

Comment: `scatter` is explained in the *Graphics Reference Manual*, but typing `scatter` y x draws a graph of y against x. The relationship, we note, appears to be nonlinear.

Note

When you draw a graph, the Graph window appears, probably covering up your Results window. Click on the **Results** button to put your Results window back on top. Want to see the graph again? Click on the **Graph** button. See Chapter 16 for more information about the **Graph** button.

Graphing data, continued

Next, we draw separate graphs for foreign and domestic cars.

```
. sort foreign
. scatter mpg weight, by(foreign, total)
```

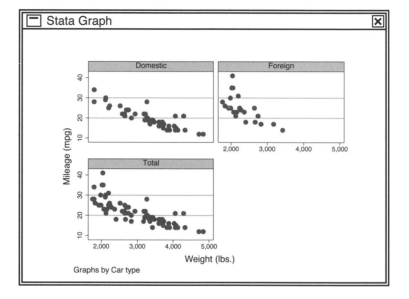

Syntax note: by() is on the right of the command; therefore, scatter did whatever it is that it does with the grouping information. What scatter did is draw separate graphs for domestic and foreign cars in a single image. We have only two groups, but scatter will allow any number — the individual graphs just get smaller. The total option added an overall graph to the image.

Analysis note: The relationship is not only nonlinear; the domestic-car relationship appears to differ from that of foreign cars.

Model fitting: linear regression

Restatement of problem: We are to model the relationship between MPG and weight.

Plan of attack: Based on the graphs, we judge the relationship nonlinear and will model MPG as a quadratic in weight. Also based on the graphs, we judge the relationship to be different for domestic and foreign cars. We will include an indicator (dummy) variable for foreign and evaluate afterwards whether this adequately describes the difference. Thus, we will fit the model:

$$\mathtt{mpg} = \beta_0 + \beta_1 \, \mathtt{weight} + \beta_2 \, \mathtt{weight}^2 + \beta_3 \, \mathtt{foreign} + \epsilon$$

foreign is already a 0/1 variable, so we only need create the weight-squared variable:

```
. gen wtsq = weight^2
. regress mpg weight wtsq foreign

      Source |       SS       df       MS              Number of obs =      74
-------------+------------------------------           F(  3,    70) =   52.25
       Model |  1689.15372      3   563.05124           Prob > F      =  0.0000
    Residual |   754.30574     70  10.7757963           R-squared     =  0.6913
-------------+------------------------------           Adj R-squared =  0.6781
       Total |  2443.45946     73  33.4720474           Root MSE      =  3.2827

------------------------------------------------------------------------------
         mpg |      Coef.   Std. Err.      t    P>|t|     [95% Conf. Interval]
-------------+----------------------------------------------------------------
      weight |  -.0165729   .0039692    -4.18   0.000    -.0244892   -.0086567
        wtsq |   1.59e-06   6.25e-07     2.55   0.013     3.45e-07    2.84e-06
     foreign |    -2.2035   1.059246    -2.08   0.041      -4.3161   -.0909002
       _cons |   56.53884   6.197383     9.12   0.000     44.17855    68.89913
------------------------------------------------------------------------------
```

Model fitting: linear regression, continued

Aside: Stata can fit many kinds of models, including logistic regression, Cox proportional hazards, etc. Click on **Help**, choose **Search...**, select **Search documentation and FAQs**, and enter `estimation` for a complete list or look up estimation in the index of the *Stata Base Reference Manual*.

Continuation of attack: We obtain the predicted values:

```
. predict mpghat
(option xb assumed; fitted values)
```

Comment: Be sure to read [U] **23 Estimation and post-estimation commands**. There are a number of features available to you after estimation — one is calculation of predicted values. `predict` just created a new variable called `mpghat` equal to

$$-.0165729\, \texttt{weight} + 1.59 \times 10^{-6}\, \texttt{wtsq} - 2.2035\, \texttt{foreign} + 56.53884$$

Model fitting: linear regression, continued

We can now graph the data and the predicted curve.

Continuation of attack: We just created `mpghat` with `predict`. We could graph the fit and data, but we want to evaluate the fit on the foreign and domestic data separately to determine if our shift parameter is adequate. Thus, we will draw the graphs separately:

```
. sort weight
. scatter mpg mpghat weight if foreign==0, msymbol(O i) connect(. l)
```

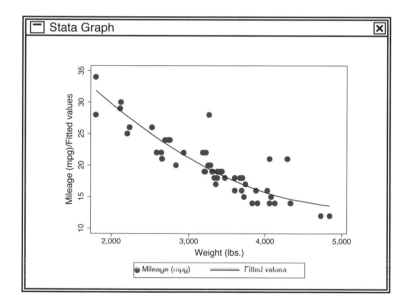

Model fitting: linear regression, continued

```
. scatter mpg mpghat weight if foreign==1, msymbol(O i) connect(. l)
```

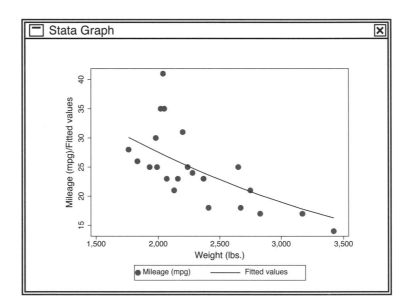

scatter mpg mpghat weight says to graph mpg versus weight and mpghat versus weight.

msymbol(O i) says use big circles for the mpg versus weight points — that is, the O (capital "oh", not a zero) — but use the invisible symbol (no symbol at all) for the mpghat versus weight points — that is, the 'i'.

connect(. l) says do not connect the mpg versus weight points — that is, the '.' — but do connect (with straight lines) the mpghat versus weight points — that is, the 'l' (ℓ). It is necessary to sort the data by the x-variable — in this case, weight — before graphing so that the points are connected in the right order.

Model fitting: linear regression, continued

Problem: You show your results to an engineer. "No," he says. "It should take twice as much energy to move 2,000 pounds 1 mile compared with moving 1,000 pounds, and therefore twice as much gasoline. Miles per gallon is not a quadratic in weight, gallons per mile is a linear function of weight."

You go back to the computer:

```
. gen gpm = 1/mpg
. label var gpm "Gallons per mile"
. sort foreign
. scatter gpm weight, by(foreign, total)
```

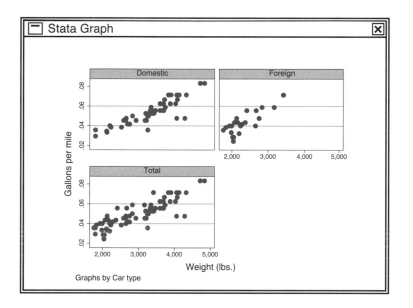

Model fitting: linear regression, continued

Satisfied the engineer is indeed correct, you rerun the regression:

```
. regress gpm weight foreign

      Source |       SS       df       MS                  Number of obs =      74
-------------+------------------------------              F(  2,    71) =  113.97
       Model |  .009117618      2  .004558809              Prob > F      =  0.0000
    Residual |   .00284001     71      .00004              R-squared     =  0.7625
-------------+------------------------------              Adj R-squared =  0.7558
       Total |  .011957628     73  .000163803              Root MSE      =  .00632

------------------------------------------------------------------------------
         gpm |      Coef.   Std. Err.       t    P>|t|     [95% Conf. Interval]
-------------+----------------------------------------------------------------
      weight |   .0000163   1.18e-06     13.74   0.000     .0000139     .0000186
     foreign |   .0062205   .0019974      3.11   0.003     .0022379     .0102032
       _cons |  -.0007348   .0040199     -0.18   0.855    -.0087504     .0072807
------------------------------------------------------------------------------
```

You find foreign cars in 1978 less efficient. Foreign cars may have yielded better gas mileage than domestic cars in 1978, but this was only because they were so light.

15 Using the Do-file Editor

The Do-file Editor

In this chapter, you will learn:

Stata has a Do-file Editor.
   You can edit do-files and other text files with it.

To enter the Do-file Editor:
   • Click on the **Do-file Editor** button
   • Or type `doedit` and press *Enter* in the Command window

The Do-file Editor lets you submit several commands to Stata at once.
   • Type the commands into the editor
   • Then press the **Do** button

The Do-file Editor has standard features found in other text editors.
   Cut, **Copy**, **Paste**, **Undo**, **Open**, **Save**, and **Print** are just a few examples.

The Do-file Editor has more advanced features to aid you in writing do-files.
   You will learn about them in this chapter.

The Do-file Editor toolbar

The Do-file Editor has eleven buttons. If you ever forget what a button does, hold the mouse pointer over a button for a moment and a box will appear with a description of that button.

New

Start a new do-file.

Open

Open a do-file from disk.

Save

Save to disk the current do-file.

Print

Print the current do-file.

Find

Search for a string in the current do-file.

Continued

The Do-file Editor toolbar, continued

Cut

Copy the selected text to the clipboard and cut it from the current do-file.

Copy

Copy the selected text to the clipboard.

Paste

Paste text from the clipboard into the current do-file.

Undo

Undo the last change.

Do

Execute the commands in the current do-file.

Run

Execute the commands in the current do-file without showing any output.

Using the Do-file Editor

Suppose that you are about to do an analysis on fuel usage for 1978 automobiles. You know that you will be issuing many commands to Stata during your analysis, and that you want to be able to reproduce your work later without having to type each of the commands again.

You may place commands in a text file; Stata can then read the file and execute each command in sequence. This file is known as a Stata "do-file"; see [U] **19 Do-files**.

Note

You can create a do-file in any text editor or word processor. You have to be careful to save the file as a text (plain ASCII) file, and you also have to watch out for the text editor placing its own extension on the file. You might tell the text editor to save `myfile.do`, but it may save `myfile.do.txt`. To force such an editor to use the extension that you want, enclose the filename in double-quotes; type `"myfile.do"` rather than `myfile.do`.

You can avoid the above hassles by using Stata's built-in Do-file Editor. Stata's Do-file Editor knows to use the `.do` extension and will automatically save the file as a text (ASCII) file.

To analyze fuel usage of 1978 automobiles, you plan to create a new variable giving gallons per mile. You want to see how that variable changes in relation to vehicle weight for both domestic and imported cars. You decide that performing a regression would be a good first step. You want to issue the following commands to Stata:

```
use http://www.stata-press.com/data/r8/auto, clear
generate gpm = 1/mpg
label var gpm "Gallons per mile"
sort foreign
regress gpm weight foreign
```

Using the Do-file Editor, continued

Click on the **Do-file Editor** button to open the Do-file Editor. Then, type the commands that you wish to submit to Stata. Note that we misspelled the variable name on the fourth line — go ahead and intentionally misspell it.

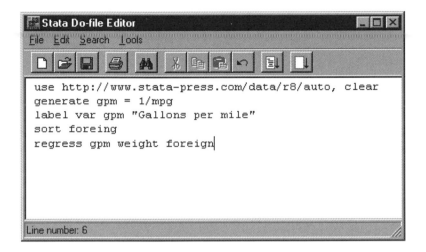

When you are through typing, press the **Do** button:

Using the Do-file Editor, continued

When you press the **Do** button, Stata executes the commands in sequence, and the results appear in the Results window:

```
. do "C:\temp\std000001.tmp"
. use http://www.stata-press.com/data/r8/auto, clear
(1978 Automobile Data)
. generate gpm = 1/mpg
. label var gpm "Gallons per mile"
. sort foreing
variable foreing not found
r(111);

end of do-file
r(111);
```

The `do "C:\...` is how Stata executes the commands that you typed in the Do-file Editor. Stata saves the commands to a temporary file and issues the `do` command to execute them.

Everything worked as planned until Stata saw the misspelled variable. The first three commands were executed, but an error was produced on the fourth. Stata does not know of a variable named `foreing`. Go back to the Do-file Editor and correct the mistake.

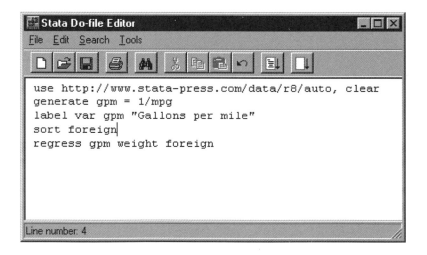

Using the Do-file Editor, continued

After you correct the spelling of `foreign` and press the **Do** button, Stata will execute each command in the editor, from the beginning.

```
. do "C:\temp\std000001.tmp"
. use http://www.stata-press.com/data/r8/auto, clear
(1978 Automobile Data)
. generate gpm = 1/mpg
. label var gpm "Gallons per mile"
. sort foreign
. regress gpm weight foreign

      Source |       SS       df       MS              Number of obs =      74
-------------+------------------------------           F(  2,    71) =  113.97
       Model |  .009117618     2   .004558809           Prob > F      =  0.0000
    Residual |   .00284001    71      .00004            R-squared     =  0.7625
-------------+------------------------------           Adj R-squared =  0.7558
       Total |  .011957628    73   .000163803           Root MSE      =  .00632

-------------+----------------------------------------------------------------
         gpm |      Coef.   Std. Err.      t    P>|t|     [95% Conf. Interval]
-------------+----------------------------------------------------------------
      weight |   .0000163   1.18e-06    13.74   0.000     .0000139    .0000186
     foreign |   .0062205   .0019974     3.11   0.003     .0022379    .0102032
       _cons |  -.0007348   .0040199    -0.18   0.855    -.0087504    .0072807
------------------------------------------------------------------------------
.
end of do-file
```

You might want to choose **Save As...** from the Do-file Editor's **File** menu and save this do-file. Later, you could **Open...** it (also from the **File** menu) and add more commands as you move forward with your analysis. By saving the commands of your analysis in a do-file as you go, you do not have to worry about retyping them with each new Stata session.

The File menu

The **File** menu of the Do-file Editor includes standard features found in most text editors. You may choose to start a **New** file, **Open...** an existing file, **Save** the current file, or save the current file under a new name with **Save As**. You may also **Print...** the current file. There are also buttons on the Do-file Editor's toolbar which correspond to these features.

There is one other useful feature under the **File** menu: you may choose **Insert File...** to insert the contents of a second file at the current cursor position in the Do-file Editor.

Editing tools

The **Edit** menu of the Do-file Editor includes the standard **Cut**, **Copy**, and **Paste** capabilities, along with a single-level **Undo**. There are also buttons on the Do-file Editor's toolbar for easy access to these capabilities. There are several other **Edit** features that you may find useful.

You may delete or select the current line.

You may also shift the current line or selection right or left one tab stop.

The **Change Case** menu item allows you to invert the case of the character to the right of the cursor or to invert an entire selection. When you **Change Case** of an entire selection, the editor will compare the number of lowercase characters with the number of uppercase characters. The case with the greater number of characters will be changed. For example, if you have aBcDe selected, you will have ABCDE after selecting **Change Case**. The editor sees that there are fewer uppercase (2) than lowercase (3) characters, and so makes the entire selection uppercase. If you select **Change Case** again, the entire selection will change to lowercase. There are fewer lowercase (0) than uppercase (5) characters, so all are switched to lowercase.

If there is no selection when you choose **Change Case**, the case of the character immediately to the right of the cursor is switched. The cursor is also moved one character to the right. For example, if the cursor is to the left of 'a' in aBcDe when you select **Change Case**, you will see ABcDe and the cursor will be just to the left of 'B'. If you select **Change Case** again, you will see AbcDe, and the cursor will be just to the left of 'c'.

Preferences

When you select **Preferences** from the **Edit** menu of the Do-file Editor, you may customize the way the editor behaves.

You may change the font and its size. You may choose any fixed-width TrueType font on your system.

You may set the number of spaces each *Tab* character represents. You may also set whether the Do-file Editor automatically indents the current line to the same level of indentation as the previous line.

Finally, you may set whether or not the Do-file Editor automatically saves the current file when you press the **Do** button or the **Run** button. Normally, the file on which you are working is not saved to disk unless you explicitly choose **Save** or **Save As** from the **File** menu. For example, if you **Open...** c:\data\myfile.do and make some changes, those changes will not be saved until you save them from the **File** menu. The changes are not saved even when you execute myfile.do by pressing the **Do** button or the **Run** button. This prevents your original file from being overwritten until you tell the Do-file Editor to overwrite it. If you check **Auto-save on Do/Run** in the **Preferences** dialog, any changes that you have made to myfile.do will be saved when you press either the **Do** button or the **Run** button.

Searching

Stata's Do-file Editor includes standard **Find...** and **Replace...** capabilities under the **Search** menu. You should already be familiar with these capabilities from other text editors and word processors. You may access the **Find** capability from a button on the toolbar as well.

The **Search** menu also includes some choices that are useful when you are writing a long do- or ado-file. You may select **Go to Line...** from the **Search** menu and jump straight to any line in your file.

Matching and balancing of parentheses ( ), braces { }, and brackets [ ] are available from the **Search** menu. When you select **Match** from the **Search** menu, the Do-file Editor looks at the character immediately to the right of the cursor. If it is one of the characters that the editor can match, the editor will find the matching character and place the cursor immediately in front of it. If there is no match, you will hear a beep and the cursor will not move.

When you select **Balance** from the **Search** menu, the Do-file Editor looks to the left and right of the current cursor position or selection, and creates a selection including the narrowest level of matching characters. If you select **Balance** again, the editor will expand the selection to include the next level of matching characters. If there is no match, you will hear a beep and the cursor will not move.

Balance is easier to explain with an example. Type (now (is the) time) in the Do-file Editor. Place the cursor between the words is and the. Select **Balance** from the **Search** menu. The Do-file Editor will select (is the). If you select **Balance** again, the Do-file Editor will select (now (is the) time).

The Tools menu

You have already learned about the **Do** button. Pressing it is equivalent to choosing **Do** from the **Tools** menu in the Do-file Editor.

Next to the **Do** button is the **Run** button. Pressing it is equivalent to choosing **Run** from the **Tools** menu. **Run** executes all the commands in the Do-file Editor just like **Do** does, but **Run** suppresses output. It is unlikely that you will ever need to use **Run**; see [U] **19.6.2 Suppressing output** for more details.

Do and **Run** are equivalent to Stata's `do` and `run` commands; see [U] **19 Do-files** for a complete discussion.

There are two other useful choices on the **Tools** menu. If you wish to execute a subset of the lines in your do-file, highlight those lines and choose **Do Selection** from the **Tools** menu.

If you wish to execute all commands from the current line through the end of the file, choose **Do to Bottom** from the **Tools** menu.

Saving interactive commands from Stata as a do-file

While working interactively with Stata, you may decide that you would like to rerun the last several commands that you typed interactively. You can save the contents of the Review window as a do-file and open that file in the Do-file Editor. See [GSW] **D.3 Saving the contents of the Review window as a do-file** for details. Also see [R] **log** for information on the `cmdlog` command, which allows you to log all commands that you type in Stata to a do-file.

16 Graphs

Working with graphs

In this chapter, you will learn

There is much to learn about Stata graphs; read the *Graphics Reference Manual* for full details.

The **Graph** button brings the Graph window to the top.

To print a graph:
- The Graph window must be open (though it need not be in front)
- Choose **Print Graph...** from the **File** menu
- Or click the **Print Graph** button

To save a graph to disk:
- Choose **Save Graph** from the **File** menu
- Enter the *filename*

If you do not specify an extension, .gph will be added and the file named *filename*.gph.

You can save the graph as a Windows Metafile (WMF) by choosing WMF as the file type in the **Save** dialog box.

To load a graph from disk:
- Type graph using *filename*

The Graph button

There is one button that allows you to quickly redisplay graphs that you have already drawn. The **Graph** button brings the Graph window to the top.

If you ever decide to close the Stata Graph window, you can only reopen it by reissuing a Stata command that draws a new graph.

Saving and printing graphs

You can save your graphs once they are displayed by choosing **Save Graph...** under the **File** menu. In the **Save** dialog box, you can choose to save your graph as either a Stata graph or a Windows Metafile (WMF). You can print the graph by choosing **Print Graph...** under the **File** menu. You can also automate the printing of many graphs. See the *Graphics Reference Manual* for a full explanation.

If you would like to copy your graph to the clipboard in order to import it into another Windows application, you should

1. Display your graph in the Stata Graph window.

2. Click on the title bar of the Stata Graph window.

3. Choose **Copy Graph** from the **Edit** menu.

4. In the other Windows application, you can then choose **Paste** from the **Edit** menu. If this option is not available, then there is most likely another method for importing objects from the clipboard. Consult the documentation for the other application to continue.

For more information on printing graphs, see the *Graphics Reference Manual*.

17 Logs: Printing and saving output

Putting it on paper and disk

In this chapter, you will learn:

A log is a recording of your Stata session.

Logs are stored as Stata formatted (SMCL) files or text (ASCII) files.
- You can view them in the Viewer
- You can print them
- You can translate formatted (SMCL) logs into text (ASCII) logs
- You can load ASCII logs into a text editor or word processor as you would any other text file

To start a log:
- Click on the **Begin Log** button and then fill in a *filename*
Or:
- Select **Log** from the **File** menu
- Click on **Begin...**
- Choose the log type
- Fill in a *filename*
Or:
- Type `log using` *filename*

The log file will be named:
- *filename*`.smcl` for SMCL logs
- *filename*`.log` for ASCII logs

If you specify a file that already exists, you will be asked to choose from:
- View the existing file
- Append the new log to the file
- Overwrite the file with the new log

The log file is closed automatically when you exit Stata.

To close the log file sooner:
- Click **Close/Suspend Log** and choose **Close log file**
- Or type: `log close`

To temporarily suspend output from being written to the log:
- Click **Close/Suspend Log** and choose **Suspend log file**
- Or type: `log off`

To resume the suspended log:
- Click **Close/Resume Log** and choose **Resume**
- Or type: `log on`

Continued

Logs

To view a snapshot of the current log in the Viewer:
- Select **Log** from the **File** menu
- Click on **View...**
- The filename of the current log is filled in automatically

Or:
- Click the **Viewer** button
- The filename of the current log is filled in automatically

To view an updated snapshot of the current log in the Viewer:
- Click on the **Refresh** button at the top of the Viewer

To print the current log:
- View a snapshot of the current log in the Viewer
- Choose **Print Viewer...** from the **File** menu

Comments can be added to your log as you work:
- In the Command window or in a do-file, type a '*' at the beginning of the line:
    ```
    * this is a comment
    * report last regression
    ```

Alternatively, you can use the `log` command to create a log file.

To start a new log file:
- Type: `log using` *filename*
- Type: `log using` *filename*, `smcl`
- Type: `log using` *filename*, `text`

To append to *filename*.`smcl` or *filename*.`log`:
- Type: `log using` *filename*, `append`

To overwrite *filename*.`smcl` or *filename*.`log`:
- Type: `log using` *filename*, `replace`

To translate a SMCL log to a text log:
- Select **Log** from the **File** menu
- Click on **Translate...**
- Enter the SMCL log file in the Input File field
- Enter the new file in the Output File field
- Click on **Translate**

To create a log that contains only the command lines that you type:
- Type: `cmdlog using` *filename*
- This log will be saved as *filename*.`txt`
- The do-file can be run to reproduce the output by typing `do` *filename*.`txt`

Logging output

All of the output that appears in the Results window can be captured in a log file. The log file can be saved as a Stata formatted (SMCL) file or as a text (plain ASCII) file. The formatted file can be printed from Stata, retaining the **bolds**, <u>underlines</u>, and *italics* that you see in the Results window. The text file can be printed from Stata or loaded into a text editor or word processor.

1. *To start a log file, click on the* **Begin Log** *button and fill in a name for the file.*
   This will open a standard file dialog box allowing you to specify a directory and filename to hold your log. If you do not specify a file extension, the extension `.smcl` will be added to the filename.

2. *If you specify a file that already exists:*
   You will be asked if you want to append the new log to the file, or else overwrite the file with the new log.

3. *A Stata formatted (SMCL) log file contains the output that you saw during your Stata session. A text (ASCII) log file will contain the ASCII translation of the output that you saw during your Stata session.*

```
. log using "C:\data\base.smcl"

      log:  C:\data\base.smcl
 log type:  smcl
opened on:   12 Sep 2002, 17:04:05
. use http://www.stata-press.com/data/r8/auto
(1978 Automobile Data)

. sort foreigh
variable foreigh not found
r(111);

. sort foreign

. by foreign: summarize price mpg


-> foreign = Domestic
    Variable |       Obs        Mean    Std. Dev.       Min        Max

       price |        52    6072.423    3097.104       3291      15906
         mpg |        52    19.82692    4.743297         12         34


-> foreign = Foreign
    Variable |       Obs        Mean    Std. Dev.       Min        Max

       price |        22    6384.682    2621.915       3748      12990
         mpg |        22    24.77273    6.611187         14         41
. * include the above means in my report
. log close
      log:  C:\data\base.smcl
 log type:  smcl
closed on:   12 Sep 2002, 17:04:13
```

4. *You can scroll through the Viewer to view previous output.*

Logging output, continued

5. *To display the log in the Viewer:* Choose **Log** from the **File** menu and select **View...**, or click on the **Viewer** button. Note: The log file image displayed in the Viewer is a snapshot taken at the time the Viewer was opened. To update the log image, click on the **Refresh** button at the top of the Viewer.

6. *You can copy from the Viewer to the clipboard and paste into the Command window:* Highlight the text that you want, pull down the **Edit** menu, and choose **Copy**. To paste into the Command window, choose **Edit—Paste**. Note: If the clipboard contains more than one line, only the first line is pasted into the Command window.

7. *You can also copy from the Viewer to the clipboard and paste into the Do-file Editor:* Highlight the text that you want, pull down the **Edit** menu, and choose **Copy**. To paste into the Do-file Editor, choose **Edit—Paste** from the menu bar of the Do-file Editor.

8. *When you copy from a* SMCL *file in the Viewer, the format of the copied material will be in plain text (*ASCII*) format.*

9. *You can add comments to your log during your Stata session.*
   If you type a '*' at the beginning of a command line, then the line is treated as a comment. Comments can be very helpful when you later read your log file and try to understand what you did (and why you did it).

```
. regress mpg weight wtsq gear_ratio displacement

  .
  . (output omitted)
  .
. regress mpg weight wtsq displacement

  .
  . (output omitted)
  .
. regress mpg weight wtsq

  .
  . (output omitted)
  .
. * the above is the regression I want to report
```

10. *The log file is closed automatically when you exit Stata.*

11. *Or you can close it whenever you want:*
    Click on the **Close/Suspend Log** button and choose **Close log file**.

12. *For more information about logs, see* [U] **18 Printing and preserving output** *and* [R] **log**. *For more information about the Viewer, see* Chapter 3.

Printing logs

1. *To print a log file from the Viewer during a Stata session:*
   Pull down the **File** menu and choose **Print Viewer...** or click on the **Print** button. After you click **OK**, an **Output Settings** dialog box will appear.

   You can fill in none, any, or all of the items "Header", "Name", and "Project". You can check or uncheck options to **Print Line #s**, **Print Header**, and **Print Logo**. These items are saved and will appear again in the **Output Settings** dialog (in this and in future Stata sessions).

   You can set the margins and color scheme that the printer will use by clicking on **Prefs...** in the **Output Settings** dialog box. Monochrome is for black-and-white printing, Color is for default color printing, and Custom 1 and Custom 2 are for customized color printing.

   You can set the font that the printer will use by clicking on **Font...** from the **Prefs** dialog box. The font dialog box will only list the fixed-width "typewriter" fonts (e.g., Courier) available for your printer. Stata, by default, will choose a font size that it thinks is appropriate for your printer.

2. *To print a closed log file:*
   To print a closed log file, you must open it first. Select **View...** from the **File** menu, enter the filename, and click **OK**. Then, print the log file according to the above instructions.

3. *If your log file is a text file (`.log` instead of `.smcl`), you can load it into a text editor (like Notepad), the Do-file Editor in Stata, or your favorite word processor.*

   From within your editor or word processor, you can edit it — add headings, comments, etc. — format it, and print it.

   Note about word processors:
   Your word processor will print out the file in its default font. The log file will not be very readable when printed in a proportionally spaced font (e.g., Times Roman or Helvetica). It will look much better printed in a fixed-width "typewriter" font (e.g., Courier).

   You may wish to associate the `.log` extension with a text editor (like Notepad, Write, or WordPad) in Windows. You can then edit and print the logs from those Windows applications if you like.

   In Windows, you associate an extension with an application by clicking once on a file with the appropriate extension and then choosing **Options** from the **View** menu. In the **Options** dialog box, click on the **File Types** tab. See the documentation that came with your operating system for more information.

   Once you have done this, any time you double-click on a file with that extension, the application associated with the extension will be opened and the file loaded into the application.

4. *If your log file is a Stata formatted file (`.smcl` instead of `.log`), you will need to open it in the Viewer to be able to see the output the way that it is displayed in the Results window.*

Rerunning commands as do-files

1. *To create a log file that contains only the command lines that you enter in a Stata session, type*

   `cmdlog using myfile`

2. *No output of any kind — no error messages, etc. — appears in a log file created with* `cmdlog`. Here's what the log shown on page 153 would look like if it had been created with the `cmdlog` option:

   ```
   use http://www.stata-press.com/data/r8/auto
   sort foreigh
   sort foreign
   by foreign: summarize price mpg
   *include the above means in my report
   ```

3. *To close the log, type*

   `cmdlog close`

4. *You can also save the contents of the Review window as a do-file.* The Review window stores the last 100 commands that you typed. You may find this a more convenient way to create a text file containing only the commands that you typed during your session. See [GSW] **D.3 Saving the contents of the Review window as a do-file** for more details.

5. *You can edit a file created with* `cmdlog` *and rerun the commands in a batch mode from within Stata.* Such files are called do-files, since the batch-mode execution is done with Stata's `do` command. Here's how to create and run a do-file:

   a. Edit the file if necessary. In the previous example, we would want to delete the mistyped command `sort foreigh`. You can open the log file in the Do-file Editor in Stata or in your favorite text editor. When the cmdlog file is closed, the command `cmdlog close` is entered, but is not included in the do-file.

   b. If you have any long command lines — ones that wrap around to two or more lines — you will have to make a few simple changes to your file. See [U] **19.1.3 Long lines in do-files** for details. If none of your command lines extend past one line, you need not be concerned about this.

   c. If you are using a word processor to edit the file, be sure to save the file as a text (plain ASCII) file. If you used the `cmdlog` option to make the file, the default extension will be `.do`.

   d. To rerun the commands, enter Stata and type do *filename*. Alternatively, you could choose **Do...** from the **File** menu and then select the file that you want to execute in the dialog.

   Note: When you run a do-file, the output will whiz past you without stopping for a —more—, so if you want to see it, you should either begin a new log or insert some `more` commands in your do-file. See [U] **10 —more— conditions**.

   e. See [U] **19 Do-files** and [U] **18 Printing and preserving output** for more information.

18 Setting font and window preferences

Changing and saving fonts and positions of your windows

In this chapter, you will learn:

You can change fonts in the following windows:

Results	(fixed-width fonts only)
Graph	(TrueType fonts only)
Viewer	(fixed-width fonts only)
Command	(any font)
Review	(any font)
Variables	(any font)
Data Editor	(fixed-width fonts only)
Do-file Editor	(fixed-width TrueType fonts only)

(Fixed-width fonts are "typewriter" fonts like Courier.)

To change the font for a window:
- Click once on the window's control-menu box (the little box in the upper-left corner of the window)
- Or right-click anywhere in the window
- Choose **Font...** and a standard Windows font dialog appears
- Select the font
- In the Do-file Editor, choose **Preferences...** from the **Edit** menu and select the font and size

You can resize and rearrange any of Stata's windows.

To save your font, graph, and window preferences:
- Pull down the **Prefs** menu and choose **Save Windowing Preferences**
- Or exit Stata; preferences will automatically be saved

The next time Stata comes up, it will come up according to these preferences.

To restore the factory settings for the fonts, graph settings, and window arrangements:
- Pull down the **Prefs** menu and choose **Default Windowing**

To go from the factory settings back to the saved preferences:
- Pull down the **Prefs** menu and choose **Load Windowing Preferences**

Note: Only one set of preferences can be saved.

Continued

Closing and opening windows

To change the color scheme of the Results window:
- Select **General Preferences...** from the **Prefs** menu
- Select the desired scheme from the **Color scheme** list

Or:
- Create your own by selecting Custom 1 from the **Color scheme** list
- Click on the various output level buttons and select the desired color
- Check the **bold** and **underline** boxes if desired

To change the color scheme of the Viewer:
- Select **General Preferences...** from the **Prefs** menu
- Click on the **Viewer Colors** tab
- Select the desired scheme from the **Color scheme** list

Or:
- Create your own by selecting Custom 1 from the **Color scheme** list
- Click on the various output level buttons and select the desired color
- Check the **bold** and **underline** boxes if desired

You can close any window except the Command and Results windows.

If you want to open a closed window or bring a hidden one to the top:
- Pull down the **Window** menu and select the desired window

There are also buttons to bring the Graph, Results, Data Editor, Do-file Editor, and Viewer windows to the top.

19 How to learn more about Stata

Where to go from here

In this chapter, you will learn:

You should now know enough to begin using Stata.

Play with Stata:
It's the best way to learn.

If you make a mistake and cannot figure out exactly
what's wrong, look at the command's help file:
Choose **Help**,
select **Stata Command...**,
and enter *commandname*.

Look at the syntax diagram and examples in the
help file and compare them with what you typed.

If you want to find the Stata command that
produces a particular statistic, try:
Choose **Help**,
select **Search...**,
select **Search documentation and FAQs**,
and enter *topic*.

Also, try the index at the end of *Base Reference
Manual*, Volume 4.

Read the:
Stata User's Guide

The *User's Guide* is designed to be read
cover to cover, and contains much useful
information about Stata.

The *Reference* manuals are not designed to be read
cover to cover, but to be sampled when needed.
For example, if you plan to do logistic
regressions, read about the `logistic` command:
[R] **logistic**

The datasets used in the examples in the *Reference*
manuals are posted at:
http://www.stata-press.com/data/

You can try performing the many examples in
these manuals using the example datasets.

Continued

Where to go from here, continued

This chapter contains some advice and
   recommended reading lists for sampling the
   User's Guide and the *Reference* manuals.

Look at the Stata web site;
   it has much useful information, including
   answers to frequently asked questions (FAQs):   *http://www.stata.com/support/faqs/*

Many useful links to Stata resources are found at: *http://www.stata.com/links/resources.html*

Do you want to learn more?
   Take a Stata NetCourse™:   NetCourse 101 is an excellent choice
   to learn about Stata. See
   http://www.stata.com/info/products/netcourse/
   for course information and schedules.

Suggested reading from the User's Guide and Reference manuals

The *User's Guide* is designed to be read from cover to cover in a linear fashion. The *Reference* manuals are designed as references to be sampled when necessary.

Ideally, after reading this *Getting Started* manual, you should read the *User's Guide* from cover to cover, but you probably want to become a Stata expert right away. Here is a suggested reading list of sections from the *User's Guide* and the *Reference* manuals. After reading and understanding these sections, you will be well on your way to being a Stata expert.

This list covers fundamental features and also points you to some less obvious features that you might otherwise overlook.

Basic elements of Stata

[U]	Chapter 14	Language syntax
[U]	Chapter 15	Data
[U]	Chapter 16	Functions and expressions

Memory

[U]	Chapter 7	Setting the size of memory
[R]	compress	Compress data in memory

Data input

[U]	Chapter 7	Setting the size of memory
[U]	Chapter 24	Commands to input data
[R]	edit	Edit and list data using Data Editor
[R]	infile	Quick reference for reading data into Stata
[R]	insheet	Read text (ASCII) data created by a spreadsheet
[R]	odbc	Load data from ODBC sources
[R]	append	Append datasets
[R]	merge	Merge datasets

Graphics

Stata Graphics Reference Manual

Useful features that you might overlook

[U]	Chapter 32	Using the Internet to keep up to date
[U]	Chapter 19	Do-files
[U]	Chapter 22	Immediate commands
[U]	Chapter 26	Commands for dealing with strings
[U]	Chapter 27	Commands for dealing with dates
[U]	Chapter 28	Commands for dealing with categorical variables
[U]	Chapter 16.5	Accessing coefficients and standard errors
[U]	Chapter 16.6	Accessing results from Stata commands

Suggested reading, continued

Estimation commands

[U] Chapter 29	Overview of Stata estimation commands
[U] Chapter 23	Estimation and post-estimation commands
[U] Chapter 16.5	Accessing coefficients and standard errors
[R] estimates	Estimation results

Basic statistics

[R] anova	Analysis of variance and covariance
[R] ci	Confidence intervals for means, proportions, and counts
[R] correlate	Correlations (covariances) of variables or estimators
[R] egen	Extensions to generate
[R] regress	Linear regression
[R] predict	Obtain predictions, residuals, etc. after estimation
[R] regression diagnostics	Regression diagnostics
[R] test	Test linear hypotheses after estimation
[R] summarize	Summary statistics
[R] table	Tables of summary statistics
[R] tabulate	One- and two-way tables of frequencies
[R] ttest	Mean comparison tests

Matrices

[U] Chapter 17	Matrix expressions
[U] Chapter 21.5	Scalars and matrices

Programming

[U] Chapter 19	Do-files
[U] Chapter 20	Ado-files
[U] Chapter 21	Programming Stata
[R] ml	Maximum-likelihood estimation

Stata Programming Reference Manual

System values

[R] set	Quick reference for system parameters
[P] creturn	Return c-class values

Internet resources

The Stata web site (*http://www.stata.com*) is a good place to get more information about Stata. Half of the web site is dedicated to user support. You will find answers to frequently asked questions (FAQs), ways to interact with other users, official Stata updates, and other useful information. You can also subscribe to Statalist, a list server devoted to Stata and statistics discussion.

In addition, you will find information on Stata NetCourses™. A NetCourse is an interactive course offered over the Internet, and varies in length from a few to eight weeks. Visit the Stata web site for more information.

At the web site is a Bookstore that contains books that we feel may be of interest to Stata users. Each book has a brief description written by a member of our technical staff, which explains why we think this book may be of interest.

If you have access to the web, we suggest that you take a quick look at the Stata web site now. You can register your copy of Stata online, and request a free subscription to the *Stata News*.

Visit *http://www.stata-press.com* for information on books, manuals, and journals published by Stata Press. The datasets used in examples in the Stata manuals are available from the Stata Press web site.

Also visit *http://www.stata-journal.com* to read about the *Stata Journal*, a quarterly publication containing articles about statistics, data analysis, teaching methods, and effective use of Stata's language.

See Chapter 20 for details on accessing official Stata updates and free additions to Stata on the Stata web site.

Notes

20 Using the Internet

Internet functionality in Stata

In this chapter, you will learn:

To use Stata to contact Stata's web site for the latest StataCorp news:
- Select **News** from the **Help** menu

To query Stata's web site to see if you have the latest official updates:
- Select **Official Updates** from the **Help** menu
- Click on `http://www.stata.com`

To use Stata to visit web sites that have new commands you can download:
- Select **SJ and User-written Programs** from the **Help** menu
- Click on `Search on Internet...`
- Or type: `search` *keyword*`, net`

To install a *Stata Journal* insert:
- Select **SJ and User-written Programs** from the **Help** menu
- Click on `http://www.stata-journal.com/software`

And then to install insert st0012 from SJ2-2
- Click on `sj2-2`
- Click on `st0012`
- Click on `click here to install`

To install an STB insert:
- Select **SJ and User-written Programs** from the **Help** menu
- Click on `http://www.stata.com`
- Click on `stb`

And then to install insert sg89 from STB-44
- Click on `stb44`
- Click on `sg89`
- Click on `click here to install`

To load a dataset that a colleague has placed on a web page:
- Type: `use http://www.stata.com/man/example.dta`

To copy a dataset over the web from a colleague:
- Type: `copy http://www.stata.com/man/example.dta mycopy.dta`

To look at a text file on a web page:
- Type: `type http://www.stata.com/man/readme.txt`

To copy a text file over the web:
- Type: `copy http://www.stata.com/man/readme.txt readme.txt`

Official Stata

By official Stata, we mean the pieces of Stata that are provided by us and supported by us. The other and equally important pieces are the user-written additions published in the STB, distributed over Statalist, or distributed in other ways.

Stata can fetch both official updates and user-written programs from the Internet. In the first case, you choose **Official Updates** from the **Help** menu. In the second case, you choose **SJ and User-written Programs** from the **Help** menu. (There are command ways of doing this, too; see [U] **32 Using the Internet to keep up to date**.)

Let's start with the official updates. There are two parts to official Stata: the Stata executable and Stata's ado-files.

StataCorp releases updates to official Stata often. These updates are to add new features and, sometimes, to fix bugs. Typically the updates are to the ado-files because, in fact, most of Stata is written in Stata's ado language. Occasionally, we update the executable as well.

When you choose **Official Updates**, Stata tells you the dates of its two official pieces:

```
update

Stata executable
     folder:                 C:\STATA\
     name of file:           wstata.exe
     currently installed:    12 Dec 2002
Ado-file updates
     folder:                 C:\STATA\ado\updates\
     names of files:         (various)
     currently installed:    12 Dec 2002
Recommendation
     compare these dates with what is available from
          http://www.stata.com
          other location of your choosing
          cdrom drive
          floppy drive
```

If you know of a location which has official updates other than the Stata web site, click on `other location of your choosing` and fill in the site or directory name that you wish.

If you are connected to the Internet, you can click on `http://www.stata.com` to compare the dates of your files with the latest updates available on the Stata web site.

Official Stata, continued

When you click on `http://www.stata.com`, Stata compares your two official pieces with the most up-do-date versions available:

```
update query

(contacting http://www.stata.com/)
Stata executable
    folder:              C:\STATA\
    name of file:        wstata.exe
    currently installed: 12 Dec 2002
    latest available:    12 Dec 2002
Ado-file updates
    folder:              C:\stata\ado\updates\
    names of file:       (various)
    currently installed: 12 Dec 2002
    latest available:    12 Dec 2002
Recommendation
    Do nothing; all files up-to-date.
```

In this case, all files are up-to-date.

We might be told that we need to update our ado-files, or update our executable, or update both. If your computer is not connected to the Internet, Stata will tell you so.

Updating the official ado-files

Much of Stata is implemented as ado-files—text files containing programs in Stata's ado language. StataCorp periodically releases updates. The **Official Updates** system makes it easy to keep your official ado-files up-to-date.

After selecting **Official Updates** from the **Help** menu and then clicking on http://www.stata.com, you might see

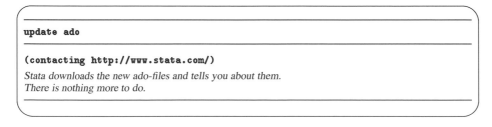

```
update query

(contacting http://www.stata.com/)
Information about other types of updates may appear . . .
Ado-file updates
    folder:                 c:\stata\ado\updates\
    names of files:         (various)
    currently installed:    12 Dec 2002
    latest available:       23 Mar 2003
Recommendation
    update ado-files
```

Stata compares the dates of your ado-files with those available from StataCorp. In this case, Stata is recommending that you update your official ado-files.

> If you do not have write permission for C:\STATA, you cannot install official updates in this way. You may still download the official updates, but you will need to use the command-line version of update; see [U] **32 Using the Internet to keep up to date** for instructions.

Click on update ado-files:

```
update ado

(contacting http://www.stata.com/)
Stata downloads the new ado-files and tells you about them.
There is nothing more to do.
```

You are done. You can skip to **Updating the executable**.

For your information, Stata does not overwrite your original ado-files, which are installed in the directory c:\stata\ado\base. Instead, Stata downloads updates to the directory c:\stata\ado\updates. If there were ever a problem with the updates, you could simply remove the files in c:\stata\ado\updates.

Updating the executable

The **Official Updates** system also checks your executable against the most current available:

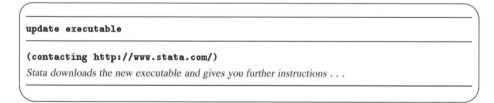

```
update query

(contacting http://www.stata.com/)
Stata executable
    folder:                 C:\STATA\
    name of file:           wstata.exe
    currently installed:    12 Dec 2002
    latest available:       03 Feb 2003
Information about other types of updates may appear . . .
Recommendation
    update executable
```

In this case, Stata recommends that you update your executable.

If you do not have write permission for C:\STATA, you cannot install official updates in this way. You may still download the official updates, but you will need to use the command-line version of update; see [U] **32 Using the Internet to keep up to date** for instructions.

Click on update executable:

```
update executable

(contacting http://www.stata.com/)
Stata downloads the new executable and gives you further instructions . . .
```

Stata copies the new executable to C:\STATA\wstata.bin; it does not replace your existing executable. That is for you to do next.

This is accomplished by simply typing update swap. Stata will then copy the newly downloaded executable over the currently running executable, and briefly restart Stata to take advantage of the new version. Once Stata has launched the new version, updating is complete, and you are free to use it as you wish.

Finding user-written programs by keyword

Stata has a built-in utility created specifically to search the Internet for user-written Stata programs. You can access it by choosing **Search...** from the **Help** menu, selecting **Search net resources**, and entering a *keyword* in the field. Equivalently, you can select **SJ and User-written Programs** from the **Help** menu, and then click on `Search on Internet...` and enter a *keyword* in the field. The utility searches all user-written programs on the Internet, including the entire collection of *Stata Journal* and STB programs. The results are displayed in the Viewer, and you can click to go to any of the matches found.

For the syntax on how to use the equivalent `search` *keywords*`,` `net` command, see [R] **search**.

Downloading user-written programs

Downloading user-written programs is easy. Start by selecting **SJ and User-written Programs** from the **Help** menu:

```
Installation and maintenance of SJ, STB, and user-written programs

    User-written programs -- SJ, STB, Statalist, and others -- are available
    from a variety of sources.  Use the links below to find, install, and
    uninstall them.

    Previously installed packages

        List
        Search...

    New packages
        Search              Search on Internet...
        Stata Journal       http://www.stata-journal.com/software
        STB and more        http://www.stata.com
        Other locations     other net sites...
                            cdrom drive
                            floppy drive

Also see
    Manual:  [U] 32 Using the Internet to keep up to date,
             [R] net
    Online:  help for _net, _search, sj, stb, update
```

What to do next will be obvious.

For instance, `lfitx2` is a command from STB-44 sg87. Your Stata has no such command:

```
. lfitx2
unrecognized command:  lfitx2
r(199);

. help lfitx2
help for lfitx2 not found
try help contents or search lfitx2
```

Downloading user-written programs, continued

You might, however, discover that you want `lfitx2` because you used Stata's `search` command and it sounded interesting.

```
. search windmeijer goodness of fit
STB-44  sg87 . . . . Windmeijer's goodness-of-fit test for logistic regression
        (help lfitx2 if installed) . . . . . . . . . . . . . . . . J. Weesie
        7/98    pp.22--27; STB Reprints Vol 8, pp.153--160
        alternative to lfit
```

Downloading user-written programs, continued

To obtain sg87 from STB-44, click on http://www.stata.com after selecting **SJ and User-written Programs** from the **Help** menu.

You should then click on stb, which will present you with a listing of all STBs available from StataCorp. Click on stb44:

```
http://www.stata.com/stb/stb44/
STB-44 July 1998

DIRECTORIES you could -net cd- to:
    ..                  Other STBs
PACKAGES you could -net describe-:
    dm59                Collapsing datasets to frequencies
    sbe19_1             Tests for publication bias in meta-analysis
    sbe24               metan -- an alternative meta-analysis command
    sg85                Moving summaries
    sg86                Continuation-ratio models for ordinal response data
    sg87                Windmeijer's goodness-of-fit test for logistic regression
    sg88                Estimating generalized ordered logit models
    sg89                Adjusted predictions and probabilities after estimation
    ssa12               Predicted survival curves for the Cox model
```

You may click on any insert to obtain more information. Click on sg87:

```
package sg87 from http://www.stata.com/stb/stb44

TITLE
      STB-44 sg87.  Windmeijer's goodness-of-fit test for logistic regression.
DESCRIPTION/AUTHOR(S)
      STB insert by Jeroen Weesie, Utrecht University, Netherlands.
      Support: weesie@weesie.fsw.ruu.nl
      After installation, see help lfitx2.
INSTALLATION FILES                              (click here to install)
      sg87/lfitx2.ado
      sg87/lfitx2.hlp
ANCILLARY FILES                                 (click here to get)
      sg87/lbw.dta
```

Click on click here to install to download and install the lfitx2.ado and lfitx2.hlp files from sg87. (If you want the lbw.dta file associated with sg87, click on click here to get.)

Downloading user-written programs, continued

You now have lfitx2. Select **Stata command** from the **Help** menu and enter lfitx2:

```
help for lfitx2                              (STB-44: sg87)

Goodness-of-fit after logistic
------------------------------
        lfitx2 [, eps(#)]
The rest of the help file is displayed . . .
```

You can install lots of commands.

You can find out what you have installed by selecting **SJ and User-written Programs** from the **Help** menu and clicking on List under Previously installed packages.

Try it after first downloading the STB insert:

```
directory of installed user-written packages

[1] package sg87 from http://www.stata.com/stb/stb44
       STB-44 sg87.  Windmeijer's goodness-of-fit test for logistic regression.
```

If you click on the one-line description of the insert, you will see the full description of the package that you installed:

```
package sg87 from http://www.stata.com/stb/stb44

TITLE                               (click here to uninstall)
       STB-44 sg87.  Windmeijer's goodness-of-fit test for logistic regression.
DESCRIPTION/AUTHOR(S)
       STB insert by Jeroen Weesie, Utrecht University, Netherlands.
       Support: weesie@weesie.fsw.ruu.nl
       After installation, see help lfitx2.
INSTALLATION FILES
       l/lfitx2.ado
       l/lfitx2.hlp
INSTALLED ON
       13 Dec 2002
```

You can uninstall materials by clicking on click here to uninstall when you are looking at the package description.

Try it.

Notes

A More on starting and stopping Stata

Contents

Below we assume that Stata is already installed. If you have not yet installed Stata, follow the installation instructions in Chapter 1.

A.1 Starting Stata

To start Stata:

1. Click on **Start**.

2. Click on **Programs**.

3. Click on **Stata**.

4. Select **Stata/SE**, **Intercooled Stata**, or **Small Stata** as appropriate; see [U] **4 Flavors of Stata**.

You will see something like this:

(Continued on next page)

175

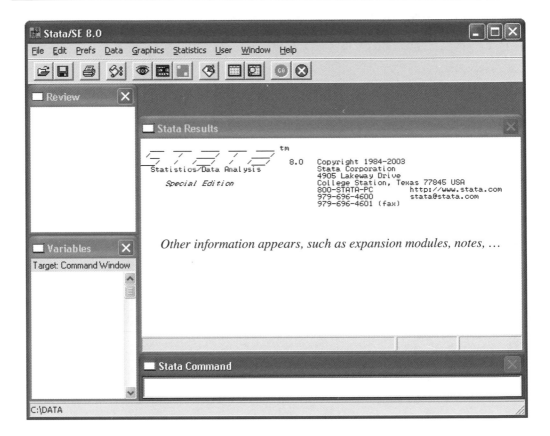

Stata is now waiting for you to type something in the Command window. If Stata does not come up, see [GSW] **B.1 If Stata does not start**.

A.2 Verifying that Stata is correctly installed

The first time that you start Stata, you should verify that it is installed correctly. Type `verinst` in the Command window, and you should see something like

```
. verinst
You are running Stata/SE 8.0 for Windows.
Stata is correctly installed.
You can type exit to exit Stata.
```

If you see an error message complaining that the command was not found, then not all of Stata is installed. Go back to Chapter 1 and reinstall Stata.

If you see any other error message, see [GSW] **B.2 verinst problems**.

Remember the `verinst` command. If you ever change your computer setup and are worried that you somehow damaged Stata in the process, you can type `verinst` and obtain the reassuring "Stata is correctly installed" message.

A.3   Exiting Stata

To exit Stata, either

1. click on the close box

    Stata works with a copy of the dataset in memory. If (1) there is a dataset in memory, and if (2) it has changed, and if (3) you click to close, a box will appear asking if it is okay to exit without saving the changes.

2. or type `exit` in the Command window.

    If you instead type `exit` and there are changed data in memory, Stata will refuse and instead say, "no; data in memory would be lost". In this case, either you must save the dataset on disk (see [R] **save**), or you can type `exit, clear` if you do not want to save the changes.

All of this is designed to prevent you from accidentally losing your data.

As you will discover, the `clear` option is allowed with all potentially destructive commands, of which `exit` is just one example. The command to bring a dataset into memory, `use`, is another example. `use` is destructive since it loads the dataset into memory and, in the process, eliminates the dataset already there.

If you type a destructive command and the dataset in memory has been safely stored on disk, Stata performs your request. If your dataset has changed in some way since it was last saved, Stata responds with the message "no; data in memory would be lost". If you want to go ahead anyway, you can retype the command and add the `clear` option. Once you become familiar with Stata's editing keys, you will discover that it is not necessary to physically retype the line. You can press the *PrevLine* key (*PgUp*) to retrieve the last command you typed and append `, clear`.

Of course, you need not wait for Stata to complain — you can add the `clear` option the first time you issue the command — if you do not mind living dangerously.

A.4   The Windows Properties Sheet

When you click on an icon to start an application in Windows, you are actually executing the instructions in that application's shortcut. The shortcut is defined on the icon's Properties Sheet.

To see the Properties Sheet for any icon, you click once on the icon to highlight it and then pull down **File** and choose **Properties**. Alternatively, you can right-click (click on the right mouse button) on the icon and then choose **Properties**.

The icons for the applications listed under the **Start** button are buried in one of these directories, depending on your configuration of Windows:

```
C:\WINDOWS\Start Menu\Programs
C:\Documents and Settings\All Users\Start Menu\Programs
C:\Documents and Settings\youruserid\Start Menu\Programs
```

Start by clicking on **My Computer** and work your way there. You will find a **Stata** folder. Click on that. You will now see icons for whatever versions of Stata you have installed — there is probably only one.

When you start Stata from the **Start** button, it is exactly the same as clicking on one of these icons.

Now that you have found the **Start** button icons, right-click on one of the Stata icons and choose **Properties**. The icon's Properties Sheet will be displayed. What you will be looking at is that icon's

General tab, and the information there is not very interesting. Click on the **Shortcut** tab. You will see something that contains the following information:

Stata/SE

Target type:	Application
Target location:	Stata
Target:	`c:\stata\wsestata.exe`
Start in:	`c:\data\`
Shortcut key:	`None`
Run:	`Normal window`

The field names may be slightly different depending on which version of Windows you are running. The names and locations of files may vary from this. For instance, Intercooled Stata and Small Stata users will see a different filename for the **Target**.

There are two things to pay attention to: the **Target** and **Start in**. **Target** is the actual command that is executed to invoke Stata. **Start in** is the directory to switch to before invoking the application.

You can change these fields and then click **OK** to save the updated Properties Sheet.

A.5   Starting Stata from other folders

You can have Stata start in whatever directory you desire. Just change the **Start in** field of Stata's Properties Sheet.

Of course, once Stata is running, you can change directories whenever you wish; see [R] **cd**.

A.6   Specifying the amount of memory allocated

You can add a /m option in the **Target** field of Stata's Properties Sheet to set the initial amount of memory allocated to Stata:

Target: `c:\stata\wsestata.exe /m20`

Only Stata/SE and Intercooled users can do this; Small Stata has no /m and the memory allocation is fixed.

The /m20 in the **Target** field tells Stata to allocate 20 megabytes of memory for data.

If you now want Stata to allocate 2 megabytes of memory, change the /m20 to /m2.

If you want 32 megabytes, change the field to /m32.

If you want 100 megabytes, change the field to /m100.

Note, if you make the amount too large, when Stata attempts to come up, you will see a message telling you that the operating system refused to provide that much memory. Stata will then attempt to come up using its default setting of 10 megabytes for Stata/SE or 1 megabyte for Intercooled Stata. If you see this message, you should decrease the amount of memory that you are asking the operating system to provide to Stata.

You can also change the amount of memory once Stata is running by using the `set memory` command, and you can make that setting permanent rather than modifying the /m option on the **Target**; see [GSW] **C. Setting the size of memory**. We recommend the `set memory` method of setting the memory allocation, as it is easier to use.

A.7 Executing commands every time Stata is started

Stata looks for the file `profile.do` when it is invoked and, if it finds it, executes the commands in it. Stata looks for `profile.do` first in the current directory, then along your PATH, then in your home directory as defined by Windows' USERPROFILE environment variable, then in the directory where Stata is installed, and finally along the adopath (see [P] **sysdir**); we recommend that you put `profile.do` in your working directory, `C:\DATA`.

Say that every time you started Stata you wanted `matsize` set to 100 (see [R] **matsize**) . Create file `C:\DATA\profile.do` containing

```
set matsize 100
```

When you invoke Stata, this command will be executed:

```
(usual opening appears, but with the addition)
running C:\DATA\profile.do ...

. _
```

Note that you could also type `set matsize 100, permanently` in Stata. The `permanently` option tells Stata to remember the setting for future sessions and eliminates the need to put the `set matsize` command in `profile.do`.

`profile.do` is treated just as any other do-file once it is executed; results are literally as if you started Stata and then typed '`run profile.do`'. The only special thing about `profile.do` is that Stata looks for it and runs it automatically.

See [U] **19 Do-files** for an explanation of do-files. They are nothing more than text (ASCII) files containing a sequence of commands for Stata to execute.

A.8 Making shortcuts

You can arrange to start Stata without going through the **Start** button. On your standard Windows screen, do you see how **My Computer** just sits on the desktop? You can put a Stata icon on the desktop, too.

The process for doing this is to find the Stata executable (not the shortcut we found above) and then drag it with the *right* mouse button where we want it. Here are the details:

1. Open the `c:\stata\` folder or whatever folder you installed Stata into.

2. In the folder, find the executable for which you want a new shortcut. The filenames are

Stata/SE:	`wsestata.exe`
Intercooled Stata:	`wstata.exe`
Small Stata:	`wsmstata.exe`

Click with the *right* mouse button and drag the appropriate executable onto the desktop.

3. When you release the mouse button, you will see a menu. Choose **Create Shortcut(s) Here**.

You have now created a shortcut. If you want the shortcut in a folder rather than on the desktop, you can drag it into whatever folder appeals to you.

You set the properties for this shortcut just as you would normally. Click with the right mouse button and choose **Properties**. You edit the Properties Sheet as explained in [GSW] **A.4 The Windows Properties Sheet**.

Note that the Properties Sheet for this new icon is different from the one for Stata under the **Start** button. The new icon could start Stata with 32 megabytes of memory, say, while the **Start** button still uses 10 megabytes.

If you create additional icons, each is different. Some users create separate icons for starting Stata with differing amounts of memory.

A.9   Executing Stata in background (batch) mode

You can run large jobs in Stata in batch mode. To do so, open a DOS window, change to your data directory, and type

```
C:\DATA> c:\stata\wsestata /b do bigjob
```

This tells Stata to execute the commands in `bigjob.do`, suppress all screen output, and route the output to `bigjob.log` in the same directory. If you desire a SMCL log file rather than an ASCII file, specify `/s` rather than `/b`.

If the do-file loads datasets that require more than the default amount of memory (10 megabytes for Stata/SE, 1 megabyte for Intercooled Stata), you will need to allocate this to Stata when you start it. Typing

```
C:\DATA> c:\stata\wstata /m15 /b do bigjob
```

will run `bigjob.do` with 15 MB of memory. Alternatively, `bigjob.do` can try to increase the memory allocated to Stata while it is running with the `set memory` command; see [GSW] **C. Setting the size of memory**.

While the do-file is executing, the Stata icon will appear on the taskbar together with a rough percentage of how much of `bigjob.do` Stata has executed. (Note that Stata calculates this percentage based on the number of characters in `bigjob.do`, so the percentage may not accurately reflect the amount of time left for the job to complete.)

If you click the icon on the taskbar, Stata will display a box asking if you want to cancel the batch job.

Once the do-file is complete, Stata will flash the icon on the taskbar on and off. You can then click the icon to close Stata. If you wish for Stata to automatically exit after running the batch do-file, use `/e` rather than `/b`.

Note that you do not have to run large do-files in batch mode. Any do-file that you run in batch mode can also be run interactively. Simply start Stata, type `log using` *filename*, and then type `do` *filename*. You can then watch the do-file running, or you can minimize Stata while the do-file is running.

A.10   Launching by double-clicking on a .dta dataset

The first time that you start Stata for Windows, Stata registers with Windows the actions to perform when you double-click on certain types of files. For example, if you double-click on a Stata dataset (`.dta` file), Stata will start and attempt to `use` the file. If you are using Stata/SE, the default amount of memory allocated to data will be 10 megabytes; for Intercooled Stata, the default amount of memory allocated to data will be 1 megabyte.

To change the default amount of memory, you may either reset the permanent default memory allocation by typing

```
. set memory #m, permanently
```

or modify the **Properties** for `.dta` datasets:

1. Double-click on **My Computer**. Pull down **View**. Select **Options**. (In some Windows versions, you may instead pull down **Tools** and select **Folder Options....**)

2. Click on the **File Types** tab. Scroll through the list until you find **Stata Dataset**. Click once on it, then click the **Edit...** button.

3. You will see an **Edit File Type** dialog. In the Actions section, click on **open** (which is the only thing there), and then click the **Edit...** button.

4. In the resulting dialog box, there will be an edit field containing something like

    ```
    c:\stata\wsestata.exe use "%1"
    ```

    This is the DOS-like command to start Stata with a dataset. Notice its similarities with the **Target** of a regular Properties Sheet. When you double-click on a Stata dataset, Windows executes this command line, replacing the %1 with the path and filename of the file on which you double-clicked.

    You can add a /m option to specify the amount of memory to allocate when double-clicking on a .dta file. If you want Stata to start with 20 megabytes of memory, add a /m20 option before the use command:

    ```
    c:\stata\wsestata.exe /m20 use "%1"
    ```

5. Once you have set /m as you wish, click **OK**, then click **Close** from the **Edit File Type** dialog. Then, click **Close** from the **Options** dialog.

Note that the /m option set here is for all .dta datasets, not a specific one. Also note that Small Stata users cannot vary the amount of memory Stata allocates to itself.

What if you want Stata to do other things as well? What if you have written a Stata program called myuse (see [U] **20 Ado-files**) that contains a list of commands that you want Stata to execute when it loads a dataset on which you double-click?

The trick is to change the command to be executed from

```
c:\stata\wsestata.exe use "%1"
```

to

```
c:\stata\wsestata.exe myuse "%1"
```

myuse.ado might read

```
                                                   ─── top of myuse.ado ───
...
if "`1'"!="" {
        use "`1'"
}
...
                                                   ─── end of myuse.ado ───
```

where ... represents the other commands in myuse.ado file. That is, we just add these three lines to the file at the point where we wish to have the dataset loaded.

A.11   Launching by double-clicking on a do-file

Double-clicking on a do-file works just like double-clicking on a dataset. When you double-click on a do-file, Stata comes up in interactive mode, executes the do-file, and then issues a prompt so that you can continue the session or exit.

You set the memory for a do-file in the same manner as you do for a .dta dataset. This time when you search for the file type, you click on **Stata do-file** rather than **Stata Dataset**.

The edit field containing the command will say do rather than use:

```
c:\stata\wsestata.exe do "%1"
```

Note that if you add a /m option here, it is for all invocations by double-clicking on a do-file, not a specific one, and it is unrelated to whatever value you may have set for double-clicking on a .dta dataset. Also note that Small Stata users cannot vary the amount of memory Stata allocates to itself.

If you click with the right mouse button on a do-file, a menu appears, and you can select **Open** or **Edit**. **Open** does the same as double-clicking on the do-file. **Edit** starts Stata, opens the Do-file Editor, and loads the do-file. You can edit the action for **Edit** in the same way as described above for the **Open** action if you wish to use an editor other than Stata's own.

A.12   Launching by double-clicking on a .gph graph file

When you double-click on a Stata .gph file, Stata comes up and displays the graph. The current directory is the directory containing the graph. Stata remains running. Type exit or click to close when you are through with the session.

When you single-click on a Stata .gph file, you can pull down **File** and select **Print** or **Print to default**. If you select **Print**, Stata comes up, displays the graph, and starts a print dialog. If you select **Print to default**, Stata comes up and prints the graph to the default printer, bypassing the print dialog. When the graph is finished printing (or when you cancel the print), Stata exits.

Equivalent to single-clicking, pulling down **File**, and choosing **Print** or **Print to default** is to right-click and choose **Print** or **Print to default**.

A.13   Running simultaneous Stata sessions

Each time you double-click on the Stata icon or launch Stata in any other way, you invoke a new instance of Stata, so if you want to run multiple Stata sessions simultaneously, you may. The title bar of each new Stata that is invoked will reflect how many copies of Stata that you are running simultaneously.

B   Troubleshooting starting and stopping Stata

Contents

B.1   If Stata does not start

You tried to start Stata and it refused; Stata or your operating system presented a message explaining that something is wrong. Here are the possibilities:

Cannot find license file

This message means just what it says; nothing is too seriously wrong, Stata simply could not find what it is looking for, probably because you did not complete the installation process or Stata is not installed where it should be.

Did you insert the codes printed on your paper license to unlock Stata? If not, go back and complete the installation; see Chapter 1.

Assuming you did unlock Stata, Stata is merely mislocated or the location has not been filled in.

Error opening or reading the file

Something is distinctly wrong and for purely technical reasons. Stata found the file that it was looking for, but either the operating system refused to let Stata open it or there was an I/O error. About the only way this could happen would be a hard-disk error. Stata technical support will be able to help you diagnose the problem; see [U] **2.9 Technical support**.

License not applicable

Stata has determined that you have a valid Stata license, but it is not applicable to the version/flavor of Stata that you are trying to run. You would get this message if, for example, you tried to run Stata for Windows using a Stata for Macintosh license.

The most common reason for this message is that you have a license for Intercooled Stata, but you are trying to run Stata/SE, or you have a license for Small Stata, but you are trying to run Intercooled Stata or Stata/SE. If this is the case, reinstall Stata, making sure to choose the appropriate flavor.

Other messages

The other messages indicate that Stata thinks you are attempting to do something that you are not licensed to do. Most commonly, you are attempting to run Stata over a network when you do not have a network license, but there are a host of other alternatives. There are two possibilities: either you really are attempting to do something that you are not licensed to do or Stata is wrong. In either case, you are going to have to call us. Your license can be upgraded, or, if Stata is wrong, we can provide codes over the telephone to make Stata stop thinking that you are violating the license. See [U] **2.9 Technical support**.

B.2   verinst problems

Once Stata is running, you can type `verinst` to check if it is correctly installed. If the installation is correct, you will see something like

```
. verinst
You are running Stata/SE 8.0 for Windows.
Stata is correctly installed.
You can type exit to exit Stata.
```

If, however, there is a problem, `verinst` will report it.

In most cases, `verinst` itself tells you what is wrong and how to fix it. There is one exception:

```
. verinst
unrecognized command
r(199);
```

This most likely means that Stata is not correctly installed. If you need to install Stata in a nonstandard way, you can learn exactly how Stata works by reading [GSW] **A. More on starting and stopping Stata**. Otherwise, reinstall Stata following the instructions in Chapter 1.

B.3   Troubleshooting

Crashes are called Application Faults in Windows. The dreaded Application Fault is typically not the application's fault. Most commonly, it is caused by configuration problems, bugs in device drivers, memory conflicts, and even hardware problems.

If you experience an Application Fault, first look at the Frequently Asked Questions (FAQ) for Windows in the user-support section of the Stata web site *http://www.stata.com*. You may find the answer to the problem there. If not, we can help, but you must give us as much information as possible.

Reboot your computer, restart Stata, and try to reproduce the fault, writing down everything you do before the fault occurs. We will want that information along with the contents of your `CONFIG.SYS`, `AUTOEXEC.BAT`, `SYSTEM.INI`, and `WIN.INI` files. If you cannot email them to us, at least print them so you can look at them when you call us.

If Stata used to work on your computer but suddenly stopped working, try to remember any hardware or software that you have recently installed.

In addition, give us as much information about your computer as possible. Is it a Pentium? A Celeron? Are you running Windows ME, 98, 95, XP, 2000, or NT? What brand is your computer? What kind of mouse do you have? What kind of video card?

Finally, we need your Stata serial number and the date your version of Stata was "born". Include them if you email, know them if you call. You can obtain them by typing `about` in Stata's Command window. `about` lets us know everything about your copy of Stata, including the version and the date it was produced.

C   Setting the size of memory

Contents

C.1   Memory size considerations

Stata works with a copy of the dataset that it loads into memory.

By default, Small Stata allocates about 300K to Stata's data areas, and you cannot change it.

By default, Intercooled Stata allocates 1 megabyte to Stata's data areas, and you can change it.

By default, Stata/SE allocates 10 megabytes to Stata's data areas, and you can change it.

You can even change the allocation to be larger than the physical amount of memory on your computer because Windows provides virtual memory.

Virtual memory is slow but adequate in rare cases when you have a dataset that is too large to load into real memory. If you use large datasets frequently, we recommend that you add more memory to your computer. See [GSW] **C.4 Virtual memory and speed considerations**.

One way to change the allocation is when you start Stata. This was discussed in [GSW] **A.6 Specifying the amount of memory allocated**.

In addition, you can change the total amount of memory allocated while Stata is running, and optionally make that setting the default to be used by future invocations of Stata. That is the topic of this chapter.

Understand that it does not much matter which method you use. Being able to change the total on the fly is convenient, but even if you cannot do this, it just means that you specify it ahead of time, and if later you need more, you must exit Stata and reinvoke it with the larger total.

C.2   Setting the size on the fly

Assume that you have changed nothing about how Stata starts, so you get the default amount of memory (10 megabytes for Stata/SE; 1 megabyte for Intercooled Stata) allocated to Stata's data areas. You are working with a large dataset and now wish to increase memory to 32 megabytes. You can type

```
. set memory 32m
```

and, if your operating system can provide the memory to Stata, Stata will work with the new total. Later in the session, if you want to release that memory and work with only 2 megabytes, you could type

```
. set memory 2m
```

There is only one restriction on the set memory command: whenever you change the total, there cannot be any data already in memory. If you have a dataset in memory, you save it, clear memory, reset the total, and then use it again. We are getting ahead of ourselves, but you might type

```
. save mydata, replace
file mydata.dta saved

. drop _all

. set memory 32m
...

. use mydata
```

When you request the new allocation, your operating system might refuse to provide it:

```
. set memory 512m
op. sys. refuses to provide memory
r(909);
```

If that happens, you are going to have to take the matter up with your operating system. In the above example, Stata asked for 512 megabytes, and the operating system said no.

C.3 The memory command

memory helps you figure out whether you have sufficient memory to do something. We are using Stata/SE, and we type memory:

```
. memory
```

	bytes	
Details of **set memory** usage		
overhead (pointers)	114,136	10.88%
data	913,088	87.08%
data + overhead	1,027,224	97.96%
free	21,344	2.04%
Total allocated	1,048,568	100.00%
Other memory usage		
set maxvar usage	1,816,666	
set matsize usage	1,315,200	
programs, saved results, etc.	772	
Total	3,132,638	
Grand total	4,181,206	

Note that the output of memory will vary slightly for Intercooled Stata; see [R] **memory**. 21,344 bytes free is not much. You might increase the amount of memory allocated to Stata's data areas by specifying set memory 2m.

```
. save nlswork
. set memory 2m
...

. use nlswork
(NLS Women 14-26 in 1968)
```

(Continued on next page)

```
. memory
```

	bytes	
Details of **set memory** usage		
overhead (pointers)	114,136	5.44%
data	913,088	43.54%
data + overhead	1,027,224	48.98%
free	1,069,920	51.02%
Total allocated	2,097,144	100.00%
Other memory usage		
set maxvar usage	1,816,666	
set matsize usage	1,315,200	
programs, saved results, etc.	773	
Total	3,132,639	
Grand total	5,229,783	

Over 1 megabyte free; that's better. See [R] **memory** for more information.

C.4   Virtual memory and speed considerations

When you open (`use`) a dataset in Stata for Windows, Stata loads the entire dataset into memory. If you have Stata/SE or Intercooled Stata, you can change the amount of memory Stata allocates; see the beginning of this chapter.

Stata can use virtual memory under Windows. This allows you to use more memory than exists on your system. Windows substitutes hard disk space in place of memory. Stata does not know when Windows does this — Stata merely asks Windows for some amount of memory, and Windows either provides it or refuses. It is entirely up to Windows whether the memory provided is real or virtual. Windows will use virtual memory when the total memory demand across all active tasks exceeds the physical total on your computer. The more memory that you allocate to Stata, the more likely Windows is to use virtual memory.

You want to avoid making Windows use virtual memory except when you really need it. Virtual memory is fine for rare situations where you must work with a dataset that will not fit into real memory (RAM), but it is not a viable alternative for daily use.

What many users do not realize is that merely allocating a lot of memory to Stata may cause Windows to start using virtual memory (paging) even if Stata is not using all that memory. From time to time we hear from users who say something like "My computer has 64 megabytes of RAM and I'm only working with a 2 megabyte dataset, but Stata is going really slow and my hard disk is spinning a lot." Invariably we discover that the user has allocated 64 MB of memory to Stata — many times more than what is necessary.

If you have a 64 MB computer and are running Windows, the maximum amount of memory that you should allocate is roughly 50 MB. Allocate more than that and Windows will start paging. The relationship between the amount of RAM and when paging begins is not linear. On a 32 MB computer, allocate more than around 22 MB and Windows will start paging. This is true even if you use small datasets. However, the probability that Windows pages on a given task is reduced with smaller datasets. As Windows XP/2000/NT users are probably already aware, Windows XP, 2000, or NT consume even more resources than Windows ME, 98, or 95, so you will experience paging even sooner under Windows XP/2000/NT.

The above assumes that Stata is the only application running. If you want to run other applications simultaneously, you should reduce the amount of memory allocated to Stata.

When you use more memory than is physically available on your computer, Stata slows down. If you are using only a little more memory than is on your computer, performance is probably not too bad. On the other hand, when you are using a lot more memory than is on your computer, performance will be noticeably affected. In these cases, we recommend that you

```
. set virtual on
```

Virtual memory systems exploit locality of reference, which means that keeping objects closer together allows virtual memory systems to run faster. set virtual controls whether Stata should perform extra work to arrange its memory to keep objects close together. By default, virtual is set off. set virtual can only be specified if you are using Stata/SE or Intercooled Stata for Windows.

In general, you want to leave set virtual set to the default of off so that Stata will run faster.

When you set virtual on, you are asking Stata to arrange its memory so that objects are kept closer together. This requires Stata to do a substantial amount of work. We recommend setting virtual on only when the amount of memory in use drastically exceeds what is physically available. In these cases, setting virtual on will help, but keep in mind that performance will still be slow. See [U] **7.5 Virtual memory and speed considerations**.

In summary, you may have to experiment to find the maximum amount of memory that you can allocate to Stata and not have Windows start paging. This is the most you will want to allocate to Stata to do your day-to-day work. Only set the amount of memory close to or above the physical amount of RAM on your computer in emergencies. For example, you may want to read in a very large dataset just to split it into pieces.

If you find that you are experiencing many "rare situations" where you must work with a dataset that will not fit in real RAM, we recommend that you purchase more memory. Memory is inexpensive, and you can generally find it cheaper from sources on the web than in catalogs. Start by searching for "memory chips" using a search engine.

D More on Stata for Windows

Contents

D.1   Using Stata datasets and graphs created on other platforms

Stata will open any Stata `.dta` dataset or `.gph` graph file regardless of the platform on which it was created, even if it was a Macintosh or Unix system. In addition, Stata for Macintosh and Stata for Unix users can use any files that you create. If you transfer a Stata file using file transfer protocol (FTP), just remember to transfer using binary mode rather than ASCII.

If you need to send a dataset to a colleague who is using Stata 7.0, first, recommend that they upgrade to the latest version of Stata. Then, save your dataset with the `saveold` command so your coleague will be able to read it even if they do not upgrade. For example,

```
saveold "c:\my data\flood2.dta"
```

D.2   Importing a Stata graph into another document

The easy way to import a Stata graph into another application is via the Windows clipboard.

Create your graph. To copy it to the clipboard, pull down **Edit–Copy Graph**. Stata will copy the graph as a Windows Metafile (WMF); this ensures that the receiving application obtains it in the highest resolution possible.

A metafile contains the commands necessary to redraw the graph. That is, a metafile is a collection of lines, points, text, and color information. Metafiles, therefore, can be edited in a structured drawing program.

After you have copied the graph to the clipboard, switch to the application into which you wish to import the graph and paste it. In most applications, this is accomplished by pulling down **Edit–Paste**. Consult the documentation for your particular application for more details.

Stata also gives you the ability to create a file containing the metafile. You could later import this file into another application. To save a graph as a metafile, pull down **File–Save Graph**. In the **Save** dialog box, choose **Windows Metafile** as the file type.

Stata can save encapsulated PostScript (EPS) files if you set your default printer driver to create EPS. EPS files can also be imported into desktop publishing applications. See the *Graphics Reference Manual* for more information.

D.3   Saving the contents of the Review window as a do-file

To save the contents of the Review window as a do-file, select **Save Review Contents** from the Review window's system menu (the little box at the left of the Review window's title bar) or right-click on the Review window and choose **Save Review Contents**. Stata will save the commands in the file you specify in the **Save** dialog. You can edit the do-file in Stata's Do-file Editor (see Chapter 15 earlier in this book), or in a text editor such as Notepad if you wish. If you use a word processor to edit the file, make sure to save the file as text. Also, be aware that some editors such as Notepad will always try to append a `.txt` extension to the filename. Even though you may have typed `myfile.do` in the **Save as** dialog box, Notepad will save `myfile.do.txt`. To force such an editor to use the extension you want, enclose the filename in double-quotes; type `"myfile.do"` rather than `myfile.do`.

For more information on do-files, see [U] **19 Do-files**.

D.4   Making Windows show file extensions

When you look at files using Windows Explorer, the file extensions may be hidden from you. This makes it difficult to rename files. For example, if you wish to rename `myfile.txt` to `myfile.do`, but extensions are hidden, you will only see `myfile` as the filename. When you click on `myfile` and type `myfile.do`, Windows will not change the extension on the file. Rather, since the `.txt` extension is hidden, you will now have a file named `myfile.do.txt`, with `myfile.do` showing and the `.txt` hidden.

You can turn off this behavior:

1. Double-click on the **My Computer** icon.

2. Pull down **Tools** and choose **Folder Options....** (In older versions of Windows, you may need to pull down **View** and choose **Options....**)

3. Click on the **View** tab in the **Options** dialog box.

4. Make sure that **Hide file extensions for known file types** is not checked. (In older versions of Windows, this may say **Hide MS-DOS file extension for file types that are registered**.)

5. Click on **OK**.

Now, you can see the full names of files in Windows Explorer. When you click on `myfile.txt` and change its name to `myfile.do`, you will not have to worry about a hidden extension remaining on the file.